STP 1344

Applications of Inductively Coupled Plasma-Mass Spectrometry to Radionuclide Determinations: Second Volume

Roy W. Morrow and Jeffrey S. Crain, Editors

ASTM Stock #: STP1344

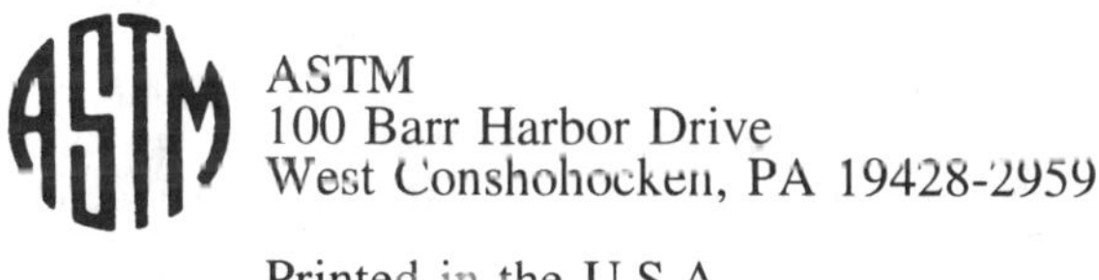

ASTM
100 Barr Harbor Drive
West Conshohocken, PA 19428-2959

Printed in the U.S.A.

Library of Congress Cataloging-in-Publication Data

The Library of Congress has cataloged
Volume I of this title as follows:

Application of inductively coupled plasma-mass spectrometry to
radionuclide determinations/ Roy W. Morrow and Jeffrey S. Crain, editors.
(STP: 1291)
"ASTM publication code number (PCN) : 04-012910-35
Application of Inductively Coupled Plasma-Mass Spectrometry to Radionuclide Determinations contains papers presented at the symposium . . . , held in Gatlinburg, Tennessee on 13-14 October 1994.
Includes bibliographical references and index.
ISBN 0-8031-2034-6
1. Radioisotopes—Analysis—Congresses. 2. Plasma spectroscopy—Congresses. 3. Mass spectroscopy—Congresses. I. Morrow, Roy W., 1942- . II. Crain, Jeffrey S., 1062- . III. Series: ASTM special technical publication: 1291.
QD466.5.A66 1995
621.48'37—dc20 95-47789
CIP

Peer Review Policy

Each paper published in this volume was evaluated by two peer reviewers and at least one editor. The authors addressed all of the reviewers' comments to the satisfaction of both the technical editor(s) and the ASTM Committee on Publications.

To make technical information available as quickly as possible, the peer-reviewed papers in this publication were prepared "camera-ready" as submitted by the authors.

The quality of the papers in this publication reflects not only the obvious efforts of the authors and the technical editor(s), but also the work of the peer reviewers. The ASTM Committee on Publications acknowledges with appreciation their dedication and contribution of time and effort on behalf of ASTM.

Printed in Chelsea, MI
October, 1998

Foreword

This publication, *Applications of Inductively Coupled Plasma-Mass Spectrometry to Radionuclide Determinations: Second Volume*, contains papers presented at "The Second Symposium on Applications of Inductively Coupled Plasma-Mass Spectrometry to Radionuclide Determinations," which was held 3-4 March, 1998 in New Orleans, Louisiana, during the 1998 Pittsburgh Conference and Exposition. The symposium was sponsored by ASTM Committee C–26 on the Nuclear Fuel Cycle and the 1998 Pittsburgh Conference Program Committee. Roy W. Morrow of Lockheed Martin Energy Systems (Oak Ridge, Tennessee) and Jeffrey S. Crain of Argonne National Laboratory (Argonne, Illinois) and Union Carbide Corporation (South Charleston, West Virginia) organized and presided over the symposium and are co-editors of this special technical publication.

Dedication—Velmer Arthur Fassel

Distinguished Professor of Chemistry, Iowa State University

The Society for Applied Spectroscopy Gold Medal Award, 1964
The Pittsburgh Spectroscopy Award, 1969 (Spectroscopy Society of Pittsburgh)
The Maurice F. Hasler Award, 1971 (Spectroscopy Society of Pittsburgh)
The American Chemical Society (ACS) Award in Analytical Chemistry, 1979
The ACS Division of Analytical Chemistry Award in Chemical Instrumentation, 1983
The Lester W. Strock Award, 1986 (Society for Applied Spectroscopy)
The ACS Division of Analytical Chemistry Award in Spectrochemical Analysis, 1988
The Pittsburgh Analytical Chemistry Award, 1995
(Society for Analytical Chemists of Pittsburgh)

On March 4, 1998, Professor Velmar Fassel passed away, leaving behind a legacy of scientific achievement which has—at the very least—shaped the development of analytical atomic spectroscopy during the latter part of this century. Starting in the 1960s, Professor Fassel and his co-workers set the course leading to development of analytical ICP emission spectroscopy, and in the 1970s, they were equally successful in developing and demonstrating the first ICP mass spectrometer. Without question, Fassel's work serves as a standard against which other analytical spectroscopists are judged, and it is likely that this will continue to be so, even in the next century.

For resons known only to a greater power, Professor Fassel's passing coincided with the symposium described in this Special Technical Publication. Without his many years of dedication and innovation, the technology underlying this symposium would not have existed; therefore, it is with great respect and appreciation that we dedicate this publication to his memory.

Contents

Overview

As the world moves inexorably toward the 21st century, the global community of scientists and engineers can look back proudly upon the achievements of this century. Diseases thought to be intractable in the latter century, e.g., polio and smallpox, were cured, and technologies that were inconceivable to the great 19th century thinkers, e.g., microprocessors and lasers, were developed. The development of controlled nuclear fission, and the ongoing efforts to harness nuclear fusion, certainly fall within the latter category; however, for reasons political and environmental, not only must we look to the future of these energy sources, we must also look at the legacy of their development.

Even in the next century, Chernobyl, Three Mile Island, and Hanford will likely serve as object lessons of nuclear technology gone awry, and it is equally likely that the technological problems created by these disasters will challenge future generations of scientists and engineers. The threat of nuclear proliferation (as exemplified by the recent Indian and Pakistani nuclear weapons tests) will also remain, as will the problems surrounding secure storage of nuclear wastes—whether of military or civilian origin. But, despite these problems, many nations will continue to rely upon nuclear energy; therefore, at the very least, development of supporting technologies must continue apace to match the level of oversight demanded by governments and the public at large.

Chemical metrology (aka analytical chemistry) will almost certainly remain as one of the key supporting technologies for nuclear energy production, waste management, and environmental restoration. A number of significant discoveries in analytical chemistry, e.g., development of solvent extractions and sulfonate cation exchange resins, came about as a consequence of early efforts to characterize nuclear materials, and as nuclear technologies have matured, so too have the supporting measurement techniques. Instruments have played a pivotal role in the maturation of these nuclear-related measurements, and this brings us to the focus of this special technical publication.

Even from its early conception [1], inductively coupled plasma-mass spectrometry (ICP-MS) was thought to be well-suited to the unique measurement problems facing the nuclear industry. These thoughts were well-founded; indeed, one might consider it unusual if a modern nuclear research center did not have access to one or more ICP mass spectrometers (quadrupole or otherwise). However, as ICP-MS has matured, improvements in sensitivity and precision have made possible measurements that were inconceivable to the ''founding fathers'' of the technology. Therefore, there is a periodic need to gather information and obatin a ''snapshot in time'' of the technology and its applications in nuclear energy.

In March of 1998, a symoposium entitled ''Applications of ICP-MS to Radionuclide Determinations'' was convened in conjunction with the 1998 Pittsburgh Conference and Exhibition to obtain the aforementioned ''snapshot in time.'' The 1998 symposium was the second of its kind—the first was held in 1994 at the ORNL/DOE Conference on Analytical Chemistry in Energy Technology [2]. However, unlike the first symposium, the second was an international event in which speakers from the United States, Europe, and the Middle East described new developments in ICP-MS relevant to the nuclear energy community. The papers presented at the 1998 symposium are published herein, and in the balance of this overview, we shall summarize the innovative techniques described by the presenters.

Applications of ICP-MS in the Nuclear Fuel Cycle

Papers by T. A. Policke et al. and J. S. Crain describe the use of ICP-MS for the assessment of isotopic composition in uranium products. The work by Policke examines the performance of quadrupole ICP-MS over a broad range of uranium enrichments, whereas Crain focuses primarily upon uranium down-blended to reactor-grade enrichment. Both papers describe methods that support uranium blending operations, and it is important to note that in the foreseeable future these operations will likely serve as a primary means of reducing highly-enriched (i.e., weapons grade) uranium inventories in the United States and elsewhere.

I. Bowen et al. describe the continued refinement of ICP multicollector MC (a magnetic sector mass spectrometer) for the precise and accurate determination of uranium isotope ratios. In their study, thorium or lead was used as an internal reference for the calibration of uranium isotope ratios. Using this method, isotope ratio precision was easily on par with thermal ionization MS (TIMS); however, sample throughput was four to five times greater than that which is typical of TIMS. This methodology may one day find equal application in characterization of nuclear and environmental media.

Papers by L. L. Tovo et al. and V. D. Jones describe applications of ICP-MS at the Savannah River Site (SRS) near Aiken, SC. Tovo and co-workers describe lessons learned while operating one of the first radiologically-contained ICP mass spectrometers (VG PlasmaQuad I, serial no. 4) in support of activities ranging from environmental monitoring to defense waste vitrification. Jones' work focuses more closely upon a specific application: characterization of high-level liquid waste. The effects of dissolved solids, which can approach 25% in the sample matrix, were studied, with particular attention paid toward nebulizer performance and ruggedness. These papers present a very clear picture of the challenges that face analytical chemists at nuclear facilities, particularly as we remediate the chemical and radiological legacy of nuclear weapons production.

Papers by P. G. Bienvenu et al. and B. Mitterrand et al. describe applications of ICP-MS toward nuclear waste management in France. Bienvenue describes the determination of long-lived β emitters (e.g., ^{135}Cs) in waste materials destined for long-term storage. Compared with traditional methods (such as liquid scintillation), ICP-MS required fewer preparative steps, thus simplifying the determination of these important radionuclides. Mitterrand describes the determination of ^{99}Tc, ^{237}Np and Th isotopes in uranyl nitrate solutions destined for reprocessing. A double-focusing, high-resolution ICP mass spectrometer was used to increase measurement sensitivity (in low resolution mode) or reduce spectral interferences (in high resolution mode). Overall, these papers present an inside look at French waste management policies, which are, in many respects, more consistent with modern environmental philosophy (e. g., "recycle and reuse") than are the United States' policies.

Applications of ICP-MS in Health and Environmental Monitoring

E. H. Evans and co-workers describe the development of on-line sample processing techniques for the determination of actinides in natural waters. With matrix elimination and analyte preconcentration, method detection limits in the pg range were achieved for Th and U; however, these were blank-limited, and yet lower limits (fg range) would be expected for anthropogenic actinides (Np, Pu, and Am). Techniques like those described by the authors underlie many ICP-MS methods for the determination of radionuclides in natural media, as exemplified by Z. Karpas et al. in their study of uranium in clinical and environmental media.

Like Karpas and co-workers, L. A. Lewis and G. K. Schweitzer, and C. J. Pickford et al. describe techniques for the determination of radionuclides in clinical media. Lewis and

Schweitzer studied the determination of ^{99}Tc in urine, and found that by using extraction chromatography for sample clean-up and analyte concentration, ICP-MS data were equally accurate and more precise than data from liquid scintillation. Pickford and co-workers obviated the need to concentrate analytes by virtue of using a double-focusing sector ICP-MS in its low resolution (high sensitivity) mode. With ultrasonic nebulization, instrument detection limits were less than 1 pg/L for 244Pu; however, several unexpected spectral interferences were encounted, and these required development of specialized preparative methods. Among other things, these papers demonstrate that, in challenging applications, even the most powerful instruments require good chemistry ''up front'' to produce meaningful data.

Papers by M. J. Ketterer and C. J. Khourey, along with R. Hearn and H. Wildner, describe ICP–MS methods for the characterization of geological media. Ketterer and Khourey used extraction chromatography and recirculating nebulization to precisely determine the $^{234}U/^{238}U$ mass ratio in natural waters and carbonate rocks. For 5 to 25 μg U, isotope ratio uncertainties were on the order of 0.2 to 0.5% RSD, which is roughly consistent with that obtained by TIMS using 100 ng U or less. In their study, Hearn and Wildner used the high sensitivity of double focusing sector ICP-MS (run in its low resolution mode) combined with ultrasonic nebulization to determine isotope ratios for several purposes, including assessment of past nuclear activity (''nuclear forensics'') and geochronometry. Both of these papers indicated that, in isotope ratio determinations, quadrupole and double-focusing sector ICP-MS are outperformed by TIMS; however, it is equally obvious that ICP-MS is more productive than TIMS. Thus, in certain applications (e.g., environmental surveys), one can expect that ICP-MS will, in the future, serve as a tool by which samples may be screened and selected for slower and more expensive TIMS measurements.

Acknowledgments

First, we wish to thank the authors, reviewers, and the ASTM Editorial staff for their tireless efforts to ensure that this Special Technical Publication would meet the high standards attained by the first volume. We also recognize the 1998 Pittsburgh Conference Committee, chaired by John Baltrus, who kindly provided travel support for our invited speakers, as well as a forum for the symposium. We thank our managers at Lockheed Martin, Argonne National Laboratory, and Union Carbide Corporation for allowing us to use company time and resources in the organization and execution of this symposium, and, finally, we thank our families for their patience and understanding over the duration of this endeavor.

Roy W. Morrow
Lockheed Martin Energy Systems,
Oak Ridge, Tennessee
Symposium Chair and Editor

Jeffrey S. Crain
Union Carbide Corporation,
South Charleston, West Virginia
Symposium Co-chair and Editor

References

[1] Houk, R. S., Fassel, V. A., Flesch, G. D., Svec, H. J., Gray, A. L., and Taylor, C. E., *Analytical Chemistry,* Vol. 52, 1980, p. 2283.

[2] *Applications of Inductively Coupled Plasma-Mass Spectrometry to Radionuclide Determinations, ASTM STP 1291,* R. W. Morrow and J. S. Crain, Eds., American Society for Testing and Materials, West Conshohocken, 1995.

Applications of ICP-MS in the Nuclear Fuel Cycle

Timothy A. Policke[1], Robert N. Bolin[2], and Terry L. Harris[3]

URANIUM ISOTOPE MEASUREMENTS BY QUADRUPOLE ICP-MS FOR PROCESS MONITORING OF ENRICHMENT

REFERENCE: Policke, T. A., Bolin, R. N., and Harris, T. L., **"Uranium Isotope Measurements by Quadrupole ICP-MS for Interim Process Monitoring of Enrichment,"** *Applications of Inductively Coupled Plasma-Mass Spectrometry to Radionuclide Determinations: Second Volume, ASTM STP 1344*, R. W. Morrow and J. S. Crain, Eds., American Society for Testing and Materials, 1998.

ABSTRACT: Historically, uranium isotopic ratio measurements in the nuclear industry have been performed using Thermal Ionization Mass Spectrometry (TIMS); primarily due to the high level of precision that can be achieved. TIMS analysis, however, requires sample purification and intricate sample loading. Quadrupole (low resolution, single detector) inductively coupled plasma - mass spectrometry, Q-ICP-MS, overcomes these disadvantages and is a cost-effective alternative, i.e., in terms of initial capital, maintenance, and operating costs. This paper presents a simple, single standard approach for measuring uranium isotope content in various solid and liquid nuclear materials along with some comparison data of Q-ICP-MS and TIMS. Intensity ratios of ^{234}U, ^{235}U, ^{236}U, and ^{238}U to total U intensity are produced, providing the enrichment level or percent ^{235}U. A detailed description of the instrument and data collection parameters is also provided. Optimal precision and accuracy are achieved through the use of a single standard which is closely matched to the enrichment and concentration of the samples. Depending upon the standard chosen, enrichments between depleted and 97% can be quantified. Standard deviations for the major uranium isotopes are typically within 0.02 absolute and at least an order of magnitude lower for the minor U isotope abundances.

KEYWORDS: ICP-MS, isotope ratio, process monitoring, quadrupole mass spectrometry

---

BWX Technologies, Inc. [formerly Babcock & Wilcox] Naval Nuclear Fuel Division (NNFD) has recently completed a campaign which employed the use of

---

[1]Principal Technical Specialist, Engineering and Quality Assurance Department (E&QAD), BWX Technologies, Inc., Lynchburg, VA

[2]Technical Specialist III, E&QAD, BWX Technologies, Inc., Lynchburg, VA.

[3]Chemistry Laboratory Manager, E&QAD, BWX Technologies, Inc., Lynchburg, VA.

Quadrupole Inductively Coupled Plasma - Mass Spectrometry (Q-ICP-MS) to monitor the enrichment downblending of high enriched recovery scrap stockpile material with low enriched uranium (U) material to produce nuclear grade reactor fuel material. The high enriched uranium (HEU) material ranged from 85% to 90% ^{235}U and existed in various forms of metal, oxides, uranium/beryllium (U/Be) rods, oxide rods, and U/Be scrap. The low enriched uranium (LEU) material (oxide) was 1.00 ± 0.02% ^{235}U. The abundance of the four major uranium isotopes, ^{234}U, ^{235}U, ^{236}U, and ^{238}U, were measured as a batch process control check of the downblended solution as well as the finished product uranyl nitrate hexahydrate (UNH) crystals. The product was certified to be acceptable under ASTM Standards and Test Methods Specification for Nuclear-Grade Uranyl Nitrate Solution (C788-93) and Specification for Uranium Hexafluoride Enriched to Less Than 5% ^{235}U (C996-90) with an enrichment of 4.00% ± 0.25% ^{235}U. For BWXT, emphasis was placed on monitoring both the ^{235}U and ^{236}U measurements. ^{236}U is an important isotope since it is a fission product and thus an indicator of the material's history. Q-ICP-MS was used to accurately determine the enrichment throughout the downblend processing. This support required 24 hour operation, seven days per week. The total amount of 4% LEU certified product was estimated at approximately 20,000 kg U or 36,000 kg with about 11,000 kg U produced to date. The average turnaround time for the downblending isotopic measurements was about 90 minutes and included a calibration standard measurement, beginning and ending control measurements, sample dilutions to match the uranium content of the samples to the standards and controls, and two to four duplicate sample measurements of five 30 second replicates for a maximum of nine sample/standard measurements in that 90 minute timespan. Typical accuracy was within 0.02% of the measured value with a percent relative standard deviation (%RSD) of less than 1% for the ^{235}U value. This precision and accuracy was acceptable by both the Nuclear Regulatory Commission (NRC) and the International Atomic Energy Agency (IAEA).

Since its commercial introduction in 1983, Q-ICP-MS has become increasingly widespread in the analytical community. Although thermal ionization mass spectrometry (TIMS) is the "classical" isotope measurement method, Q-ICP-MS is proving to be an acceptable and cost-effective alternative. The advantages of Q-ICP-MS stem from its ease of operation: the instrument operates under atmospheric pressure, samples are introduced as aqueous solutions, chemically pure separations are not necessary (such as anion exchange extraction for radiochemistry), operator skill level is less, and instrumentation is simpler. This ease leads to a much higher throughput. The advantage of TIMS is its excellent precision and accuracy of less than 0.05% (percent standard error) [*1,2*]. Whereas the short-term precision of isotope ratios by Q-ICP-MS has typically been limited to 0.2 to 0.6 % RSD for 10 replicates [*3*]. A factor attributing to achieving better precision is the count rate[*3,4*]. Although Vanhaecke, et. al., for High Resolution Inductively Coupled Plasma - Mass Spectrometry (HR-ICP-MS), documented the improvements of measuring count rates greater than 100,000 counts per second (cps), for the quadrupole units at BWXT, optimal precision was found at a count rate of about 2 million cps.

Experimental Method

Instrumentation

Two Q-ICP-MS units supported the downblending production process. A VG Elemental PlasmaQuad PQ2+ upgraded with magnetic levetating turbomolecular pumps (TMPs) and a high performance interface (HPI) predominantly supported the nominal 4% isotopic processing measurements and the HEU feedstock isotopic measurements. A Fisons Instruments Elemental Analysis PQS with ceramic TMPs provided backup support for the 4% isotopic measurements, provided the 1% blend stock material isotopic measurements, and performed the impurity measurements for product certification. Both instruments were equipped with a quartz double pass Scott spray chamber (cooled to 5°C), 1.5 mm quartz torch, quartz elbow, and 1 mL/min quartz concentric nebulizer. The instrument operating conditions and data acquisition parameters are listed in Table 1.

TABLE 1 -- *Q-ICP-MS operating conditions.*

Plasma frequency	27.12 MHz
Forward/Reflected r.f. power	nominal 1350/<5 W
Coolant/Auxiliary/Nebulizer flow rates	nominal 14/1/0.9 L min^{-1}
Nebulizer back pressure	nominal 30 psi
Expansion pressure	nominal 2 mbar
Analyzer pressure	nominal 2 x 10^{-6} mbar
Solution sample flow rate	natural uptake
Pumped waste flow rate	nominal 2.5 mL min^{-1}
Sampler cone	nickel, 1 mm aperture
Skimmer cone	nickel, 0.75 mm aperture
Data collecting mode	peak jump
Number of channels/DAC steps	7/1
Dwell time - U233, U234, U236	81.96 μsec
Dwell time - U235 (HEU)	10.24 μsec
Dwell time - U235 (LEU)	40.96 μsec
Dwell time - U238 (HEU)	40.96 μsec
Dwell time - U238 (LEU)	10.24 μsec
Optional background/oxide masses	232, 237, 239, 250, 251, 254 amu
Background/oxide dwell time	10.24 msec
Rest mass	220 amu
Acquisition time	15 or 30 seconds
Number of replicates/duplicates	at least 3/at least 2
Total number averaged	at least 8
Detector mode	Pulse Counting
Ion lens tuning	^{235}U or ^{238}U
Quadrupole settle time	2 - 5 ms
Deadtime correction factor	15 - 25

Calibration and Measurement

A peak resolution check was performed daily using ^{235}U or ^{238}U. The resolution, full-width-tenth-maximum (FWTM) peak height, was targeted for 0.70 $\pm$ 0.15 amu. The instrument was conditioned for 1 to 2 hours with the enrichment standard solution which is to be analyzed. An instrument stability and sensitivity check was typically performed. Any changes in enrichments to be analyzed require that the instrument system be flushed thoroughly with a 4% HNO_3/0.5% HF acid rinse solution before the conditioning step; particularly when changing from HEU to LEU or vice versa. The instrument glassware and cones can be changed when changing enrichment levels to help facilitate this transition.

For the nominal 4% ^{235}U data, calibration was performed using New Brunswick Laboratory (NBL) Certified Reference Material (CRM) U050. The isotopic calibration was verified with measurements of both CRM U050 (separate dilution from stock) and CRM U030, as controls. Both of these independent U030 and U050 measurements were required to meet BWXT's nuclear material control program statistical limits, 0.04% and 0.02%, respectively, providing an indication of the linearity of the use of the single standard, U050. For the nominal 90% ^{235}U data, calibration was performed using NBL CRM U900. The calibration was verified with another measurement (separate dilution from stock) of CRM U900, as a control. Periodically, BWXT verified the calibration using CRM U930, as a control, of the CRM U900 data. For the nominal 97% ^{235}U data, calibration was performed using NBL CRM U970 and verified with another measurement (separate dilution) of CRM U970, as a control. The certified values are tabulated in Table 2. Besides ^{235}U, both ^{234}U and ^{236}U were monitored for accuracy.

TABLE 2 -- *Certified weight percent values for NBL CRM material.*

Isotope	U030	U030A	U050	U900	U930	U970
^{234}U	0.0187	0.02732	0.0275	0.7735	1.0759	1.6582
^{235}U	3.009	3.0032	4.949	90.098	93.276	97.663
^{236}U	0.0202	0.000594	0.0476	0.3337	0.2034	0.1497
^{238}U	96.953	96.9689	94.975	8.795	5.445	0.5296

The standard instrument software was used to generate the mass bias coefficients for each uranium isotope used to calculate the isotope abundance information. The interference equation capability of the PQ software was used to sum the contributions of the individual uranium isotopes and store that sum as ^{238}U for LEU. The standard ratio feature of the software calculated the abundance of ^{234}U as $^{234}U/^{238}U$ (total U) and similarly the abundances of ^{235}U and ^{236}U. The actual abundance of ^{238}U was determined by difference. (For HEU, the isotopes are summed in ^{235}U, the individual abundances are calculated with this mass in the denominator, and the ^{235}U is obtained by difference.) The U content of the final solutions that were analyzed was typically 100 to 250 ppb, depending upon instrument sensitivity at the time of measurement. The count rate for total uranium content was optimally found to be 2 million cps, as shown in Table 3 (where U238* is the sum total of all U isotopes) and displayed in Figure 1, noting the plateau region. Note that the emphasis was not on the concentration but on the count rate.

TABLE 3 -- *2 million counts per second data.*

Cps * 10E-6	U235/U238* Ratio
0.105	5.491
0.217	5.383
0.418	5.282
0.659	5.223
0.866	5.167
1.127	5.128
1.343	5.074
1.591	5.039
1.832	4.962
2.083	4.949
2.316	4.958
2.604	4.919
2.831	4.875
3.066	4.810
3.454	4.800

Cps = counts per second.
U238* is the sum of all of the U isotopes.

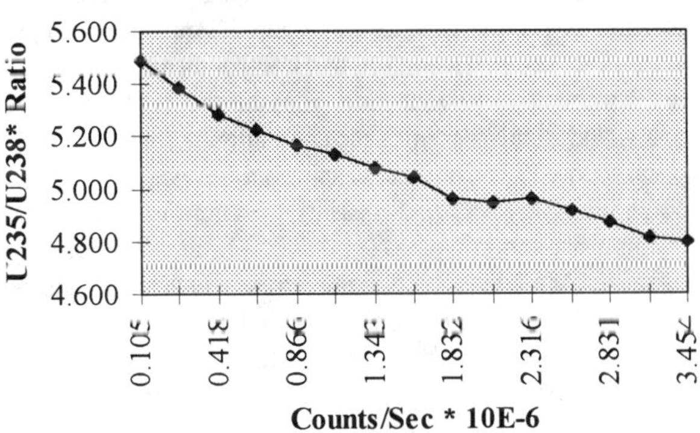

FIG. 1 – *2 million counts per second determination.*

Reagents

Seastar double distilled, high purity nitric acid (HNO_3) and high purity hydrofluoric acid (HF) were used to dissolve the uranium material. Since the objective is to determine uranium isotopics only, reagent grade HNO_3 and HF may be satisfactory.

Criterion Scientific ASTM Type I water (dH_2O) purchased from Baxter/VWR or nominal 18 mohm water was used to prepare the dilutions. Normal laboratory deionized or demineralized or distilled water may be acceptable.

Nitric/Hydrofluoric Acid Rinse Solution consisting of 4% HNO_3 v/v and 0.5% HF v/v was employed to clean the sample introduction system between enrichment conditionings, i.e., from HEU to LEU. If the rinsing proved unsuccessful, then the sample introduction system was replaced with another acid cleaned system.

The following U_3O_8 New Brunswick Laboratory (NBL), Certified Reference Materials (CRMs) were used to calibrate and operate as controls for the isotope ratios: U970, U930, U900, U050, U030. Other CRMs with different enrichment levels can be used as long as the enrichment approximates that in the samples. These CRMs were dissolved using HNO_3 and the stock dissolution diluted and in nominal 10% HNO_3. Initial working stock solutions were diluted in nominal 5% acid solution and final measuring solutions were diluted with dH_2O. Since the calibration utilizes only a single standard, the isotopic calibration standard and controls should be within 1 to 2 Wt% of the ^{235}U enrichment to be analyzed in unknown sample materials. Likewise, the low abundance isotopes (^{234}U and ^{236}U) should be of similar magnitude between standards and controls. It is recommended that separate primary and secondary stock solutions be made from a separate and preferably independent source of enrichment standard.

Procedure

A sample of the nuclear-grade material (nominally 0.2 to 0.5 g) was digested in HNO_3 acid or a HNO_3/HF acid mixture and diluted in series to a concentration of approximately 100 to 250 ng/g of U. Other dissolution methods may be used. Either natural uptake, as employed for these measurements, or a standard peristaltic pump can be used for sample introduction into the plasma. The uranium intensity (i.e., concentration), as initially indicated by a ratemeter reading, was adjusted to within 10% tolerance of the targeted 2 million cps to provide good precision and reduced bias for all sample, standard, and control measurements. The tighter the control of this tolerance the more assurity of accurate measurements, due to matching of samples (including controls) and standards, since the calibration is based on the use of a single standard. A calibration standard was run and all sample analyses were bracketed by controls. The software and run conditions were set up to measure the intensity ratios of each low abundance isotope to the intensity sum of ^{233}U, ^{234}U, ^{235}U, ^{236}U, and ^{238}U, obtaining in actuality the abundance of each isotope. Mass bias correction factors, which are established using the instrument software and the calibration data, are then applied to the sample and control data. The corrected ratio measurement for a low abundance isotope is equal to the abundance of that isotope (for example, the ^{234}U intensity/U isotope intensity sum equals the ^{234}U abundance). The high abundance isotope is determined by subtracting the low abundances from 100%. Again, ^{233}U is insignificant and could be removed from the calculation.

The instrument was started up with standard recommended practices and conditioned for the isotopic enrichment of interest. Based on the ratemeter response (set to monitor the most abundant U isotope), the U content in the standard solution was adjusted to achieve the desired intensity, e.g., 2 million cps, allowing for the contributions of the minor isotopes. Sufficient uptake time was allowed before acquiring the isotope data. A rinse was not applied between standards, controls, or samples so that the signal conditioning and the conditioning of the sample in the sample introduction system was only minimally changed. The uptake line was moved directly between each sample to be analyzed and sufficient time was given for the signal to stabilize. The isotopic calibration standard was used to establish the mass bias factors. The isotopic control was used to ensure stability of the mass bias factors. Samples were bracketed between controls. Each sample was run in duplicate with the average reported. Ideally, the duplicates should fall above and below the targeted intensity range to ensure the calibration is independent of or has minimal effect due to concentration. Samples were diluted to obtain the targeted intensity level.

It is fairly easily achievable that intensity measurements for an isotope above 1% abundance to have a %RSD of less than 3% and ratio measurements to have a RSD of less than 1% for the 5 trials. Accuracy and precision can be improved through averaging, including increasing the number of trials and/or averaging replicate measurements. For replicate measurements, the difference in the high abundance isotope should be less than 0.05 Wt%.

Calculations

Two separate equations (interference equations in the PQ software) are set up for use in the data manipulation, one for high enriched isotopic measurements and one for low enriched analyses. In each case, the high abundance isotope intensity is set to equal the sum of all of the U isotope intensities. The equations are shown below. The sum of the intensities are defined as either $^{235}U^*$ or $^{238}U^*$, for high enrichment and low enrichment analyses, respectively.

HEU enrichment analysis equation:

$$^{235}U^* = {}^{233}Uint + {}^{234}Uint + {}^{235}Uint + {}^{236}Uint + {}^{238}Uint$$

LEU enrichment analysis equation:

$$^{238}U^* = {}^{233}Uint + {}^{234}Uint + {}^{235}Uint + {}^{236}Uint + {}^{238}Uint$$

where

$^{233}Uint$ = the measured ^{233}U peak intensity,
$^{234}Uint$ = the measured ^{234}U peak intensity,
$^{235}Uint$ = the measured ^{235}U peak intensity,
$^{236}Uint$ = the measured ^{236}U peak intensity,
$^{238}Uint$ = the measured ^{238}U peak intensity.

Although the intensity sums are defined with the (*) notation above, when the interference equation is applied in the software, the instrument computer generated data will give the ^{235}U intensity as equal to $^{235}U^*$ and ^{238}U intensity as equal to $^{238}U^*$ for the high enrichment and low enrichment analyses, respectively.

A calibration standard file is set up for the NBL CRM which is to be used. The isotopic Wt% of each isotope is entered into a calibration standard file as a ratio of each low abundance isotope to the intensity sum defined above. Refer to the equations below:

HEU calibration standard file, with the equation applied is equal to the software entry:

Wt% $^{233}U = {}^{233}Uint/{}^{235}U^*$	"$^{233}U/^{235}U$"
Wt% $^{234}U = {}^{234}Uint/{}^{235}U^*$	"$^{234}U/^{235}U$"
Wt% $^{236}U = {}^{236}Uint/{}^{235}U^*$	"$^{236}U/^{235}U$"
Wt% $^{238}U = {}^{238}Uint/{}^{235}U^*$	"$^{238}U/^{235}U$"

LEU calibration standard file, with the equation applied, is equal to the software entry:

Wt% $^{233}U = {}^{233}Uint/{}^{238}U^*$	"$^{233}U/^{238}U$"
Wt% $^{234}U = {}^{234}Uint/{}^{238}U^*$	"$^{234}U/^{238}U$"
Wt% $^{235}U = {}^{235}Uint/{}^{238}U^*$	"$^{235}U/^{238}U$"
Wt% $^{236}U = {}^{236}Uint/{}^{238}U^*$	"$^{236}U/^{238}U$"

The instrument software applies the appropriate equation and represents the sum of the uranium isotopes as either ^{235}U or ^{238}U. The software calculates bias factors for each of the measured ratios. The software uses the calibration standard analysis data and the calibration standard file. Any factor which is outside of the normal trended range may indicate a system contamination problem from different enrichment materials recently run and would indicate the need to clean the system. The low abundance isotopic enrichments are obtained directly from the measurement data. The high abundance isotopic enrichment is calculated by difference as follows:

HEU enrichment analysis data:

$$\text{Wt\% } ^{235}U = 100\% - \text{Wt\% } ^{233}U - \text{Wt\% } ^{234}U - \text{Wt\% } ^{236}U - \text{Wt\% } ^{238}U$$

LEU enrichment analysis data:

$$\text{Wt\% } ^{238}U = 100\% - \text{Wt\% } ^{233}U - \text{Wt\% } ^{234}U - \text{Wt\% } ^{235}U - \text{Wt\% } ^{236}U$$

Blank or background correction is not utilized. It is intended that the count rate for the minor isotopes is sufficiently greater than their background that this correction is negligible. Experimentation has proven that to be correct for the most part; however, CRM U030-A which has a ^{236}U content of 0.000594% is not accurately achievable. CRM U030 has a ^{236}U content of 0.0202 and thus was used for the low enriched uranium results presented here instead of CRM U030-A.

Results

Table 4 contains data from 11 individual and distinct lots of nominal 90% and 97% materials. Both Q-ICP-MS and TIMS were performed on those lots and there is good agreement between the two techniques with differences ranging from -0.023 to 0.032, providing an indication of the accuracy of Q-ICP-MS. Table 5 shows the Percent Relative Difference (%RD) of Q-ICP-MS measured values and New Brunswick Laboratory's certified TIMS values of material used in their Safeguards Measurement Evaluation Program [5]. Since the Q-ICP-MS measurements were performed on two different days, noted as A and B, that data provides both an indication of accuracy and precision. Control data for the 3%, 5%, 90%, 93%, and 97% data are provided in Table 6. Table 7 shows some actual 4% product data for 10 lots of downblended material, providing an indication of the accuracy and precision achievable by liquid blending of HEU and LEU materials. The product material is well within the allowable 0.25% and is actually within 0.1%. This data also shows the value of Q-ICP-MS for the process monitoring of enrichment and its role in confirming the engineering calculations.

TABLE 4 -- *Nominal 90% and 97% data.*

ID	TIMS*	Q-ICP-MS	Diff	ID	TIMS*	Q-ICP-MS	Diff
1	89.495	89.516	-0.021	1	97.275	97.284	-0.009
2	89.970	89.955	0.015	2	97.270	97.254	0.016
3	89.724	89.698	0.026	3	97.276	97.262	0.014
4	89.927	89.950	-0.023	4	97.318	97.338	-0.020
5	89.394	89.389	0.005	5	97.308	97.306	0.002
6	89.339	89.351	-0.012	6	97.386	97.395	-0.009
7	89.420	89.398	0.022	7	97.435	97.403	0.032
8	89.323	89.301	0.022	8	97.511	97.496	0.015
9	89.325	89.308	0.017	9	97.543	97.520	0.023
10	89.401	89.423	-0.022	10	97.608	97.620	-0.012
11	89.444	89.460	-0.016	11	97.611	97.618	-0.007

* Measurements performed by an independent subcontract laboratory.

Discussion

The ^{233}U isotope is primarily measured to qualitatively determine its presence by comparing the ^{233}U peak intensity to a background point since it is not normally found present in materials. The data presented here does not contain any ^{233}U data.

A single standard calibration technique is used. Optimal accuracy (or a small bias) is achieved through the use of a single standard which is closely matched to the enrichment of the samples. The intensity or concentration is also adjusted to within 10% tolerance range of 2 million cps to provide good statistical counting precision for the low abundance isotopes while maintaining a small bias for the high abundance isotopes, resulting from high intensity deadtime effects. Depending upon the standard(s) chosen, enrichments between depleted and 97% can be quantified, using a closely matched standard. The calibration and measurements are madeby measuring the intensity ratios of each low abundance isotope to the intensity sum of ^{234}U, ^{235}U, ^{236}U, and ^{238}U. The high abundance isotope is obtained by difference. The instrument is calibrated and the samples measured in units of isotopic weight percent (Wt%).

TABLE 5 -- *Safeguards Measurement Evaluation Program -- data evaluation.*

Sample Number	Analysis Wt% ^{235}U	% Relative Difference
41-42 A	89.6923	0.0150
41-44 A	89.6662	-0.0141
43-40 A	90.3385	0.0015
43-41 A	90.3216	-0.0172
41-42 B	89.6801	0.0014
41-44 B	89.6677	-0.0124
43-40 B	90.3436	0.0071
43-41 B	90.3349	-0.0025

Mean % Difference: -0.003
95% C. L. of Mean (df=7): 0.009

Mean Absolute % Difference 0.009
Standard Deviation: 0.011

% Relative Difference is the difference between the certified value (TIMS [*6*]) and the measured value (Q-ICP-MS).

TABLE 6 – *^{235}U isotopic control data for 3%, 5%, 90%, 93%, and 97% CRMs.*

	U030	U050	U900	U930	U970
	3.001	4.953	90.102	93.268	97.665
	2.996	4.940	90.124	93.281	97.666
	3.018	4.937	90.115	93.251	97.684
	3.010	4.961	90.055	93.263	97.653
	3.009	4.956	90.112	93.298	97.659
	3.027	4.942	90.118	93.261	97.651
	3.002	4.934	90.076	93.300	97.660
	3.027	4.961	90.111	93.294	97.659
	3.001	4.948	90.089	93.252	97.655
	3.015	4.956	90.099	93.279	97.668
Avg	3.011	4.949	90.100	93.275	97.662
Std Dev	0.011	0.010	0.021	0.018	0.010
%RSD	0.365%	0.202%	0.023%	0.019%	0.010%
Certified Value	3.009	4.949	90.098	93.276	97.663

TABLE 7 -- *4% data of Product Material as measured by Q-ICP-MS.*

ID	Q-ICP-MS
1	4.098
2	4.052
3	3.932
4	4.026
5	3.938
6	3.898
7	4.053
8	4.080
9	4.017
10	3.994

There are a number of factors that affect isotope ratio measurements. These include instrument tuning of the lenses (extractor, collector, lenses 1 - 3, pole bias), resolution, delta mass, dead time, amplifier offset, discriminator setting, number of points per peak (with DAC step designation), and detector voltage (pulse counting mode for data presented here). All of these factors have an affect on the peak shape, position, and calibration and thus have an affect on the mass bias factors. Reproducibility of these bias factors is paramount to obtaining accurate isotope ratio data. This reproducibility is best achieved when the peak shapes are symmetrical with respect to the centroid of the mass designation and thus produce, in peak jumping mode, flat peaks. Tolerance for the peak flatness was limited to about 2% of the major isotope's peak height.

There are two basic approaches to performing isotope ratios using Q-ICP-MS which has only one detector. The historical first approach is to set the dwell times for each mass in inverse proportion to their abundance, i.e., for low abundance masses use larger dwell times and for high abundance masses use small dwell times. With the success of multi-collector (MC) technology and the experience of TIMS, the second approach is to simulate a MC system by setting dwell times to minimal values, allowing rapid scanning with a large number of sweeps. Since the mass window, ^{234}U through ^{238}U, is small, the quadrupole settle time is virtually negligible and can be minimized. BWXT experimented with both approaches and determined that for their units the first approach produced better precision. Bonnets, although they reduce plasma flicker noise [7], were not employed for these measurements. Mass 220 was chosen as a rest mass since it is very close to the measurement masses and is a very good background mass.

Interferences can occur from adjacent isotopes of high concentration, such as an intense ^{235}U peak interfering with the measurement of ^{234}U and ^{236}U. This is particularly the case for instruments that provide only nominal unit mass resolution at 10% of the peak height. For the measurements presented here, the Q-ICP-MS peak resolution for ^{235}U was set to within 0.70 $\pm$ 0.15 amu FWTM peak height to reduce adjacent peak interference effects. ^{235}U could potentially interfere with ^{236}U determinations by forming a UH^+ ion. Hydride, and other molecular ion formations such as oxides, can be minimized by optimizing the gas flow rates (especially nebulizer) and forward r.f. power [*8*], torch box positioning, and spray chamber temperature. The oxide formations, which can be monitored at mass/charge ratios of 250, 251, 252, and 254, only affect the intensity of the uranium signal and does not cause an overlying mass interference. The use of a calibration standard which is similar in isotopic composition and intensity to the samples reduces the potential bias from this interference effect. The bias from the UH^+ interference only becomes significant for the integrated peak intensity of ^{236}U when the sample intensity deviates from the calibration standard intensity and it is very low, i.e., near the background intensity contribution. A naturally enriched standard, or one such as CRM U030-A, which contains only 0.000594% ^{236}U, can be used to test the significance of the $^{235}U\text{-}H^+$ interference on ^{236}U. Alternatively, one can monitor, for potential hydride interference, $^{238}U\text{-}H^+$ at mass/charge ratio of 239. Memory effects or sample carryover can occur from previously run samples. These effects can be detected by looking at the standard deviation of the repeat trials from a sample analysis and whether the peak intensities are random between the repeat trials or whether they drift toward increasing or decreasing intensity. Also, the percent standard deviation (% SD) of the intensity ratios should be less than or on the same order of the % SD of the peak intensities. If it is much higher, then it may be an indication of a memory effect from a sample containing a different enrichment level. It could also be indicative of general instrument instability or sample uptake and delivery to the plasma.

Conclusion

In conclusion, Q-ICP-MS is a satisfactory tool for performing uranium isotopic measurements, especially in cases where turnaround time and minimal sample preparation dominate over explicit high precision and high accuracy. For the process monitoring of downblending operations, Q-ICP-MS turned out to be appropriate and an acceptable measurement method by the governing agencies: the Nuclear Regulatory Commission, the International Atomic Energy Agency, and BWXT's nuclear material control personnel. This application of Q-ICP-MS demonstrated its advantages over other techniques, such as TIMS and radiochemistry. Those advantages include single standard calibration, minimal sample preparation, minimal corrections, and minimal personnel skill level with an average turnaround time of about 1.5 hours from calibration through the last control measurement and result reporting.

References

[*1*] Potter, D. and Carey, G., "Uranium Isotope Ratios and Detection Limits by ICP-MS," Hewlett-Packard Application Note 228-317, March 1995.

[*2*] Walder, A. J. and Hodgson, T., "The Isotopic Ratio Measurement of Uranium in the Form of Hydrolyzed Uranium Hexafluoride by Inductively Coupled Plasma Multiple Collector Mass Spectrometry," *Applications of Inductively Coupled Plasma-Mass Spectrometry to Radionuclide Determinations, ASTM STP 1291*, R. W. Morrow and J. S. Crain, Eds., American Society for Testing and Materials, 1995, pp. 20 - 25.

[3] Vanhaecke, F., Moens, L., Dams, R., and Taylor, P., Anal. Chem., 1996, 68(3), pp. 567 - 569.

[4] Vanhaecke, F., Moens, L., Dams, R., Papadakis, I., and Taylor, P., Anal. Chem., 1997, 69(2), pp. 268-273.

[5] Spaletto, M. I., "Subject: New Brunswick Laboratory Safeguards Measurement Evaluation Program, Uranium Data Evaluation Report," U. S. Department of Energy, New Brunswick Laboratory, Letter dated August 5, 1997.

[6] Spaletto, M. I., "Safeguards Measurement Evaluation Program: Uranium Sample Exchange and Plutonium Sample Exchange," U. S. Department of Energy, New Brunswick Laboratory, NBL-340, January 1998.

[7] Liezers, M., "High Precision and Accurate Lead Isotope Ratio Measurement by Quadrupole Inductively Coupled Plasma Mass Spectrometery ICP-MS," Federation of Analytical Chemists and Spectroscopy Societies - Poster Session, 1996.

[8] Jarvis, K.E., Gray, A.L., and Houk, R.S., *Handbook of Inductively Coupled Plasma Mass Spectrometry*, Blackie and Son Ltd., Glasgow and London, or Chapman and Hall, New York, 1992.

Jeffrey S. Crain[1]

A STUDY OF ISOTOPIC HOMOGENEITY IN "BLENDED URANIUM" USING ICP-MS

REFERENCE: Crain, J.S., **"A Study Of Isotopic Homogeneity In "Blended Uranium" Using ICP-MS,"** *Applications of Inductively Coupled Plasma-Mass Spectrometry to Radionuclide Determinations: Second Volume, ASTM STP 1344*, R.W. Morrow and J.S. Crain, Eds., American Society for Testing and Materials, 1998.

ABSTRACT: A 20 kg uranium metal ingot was prepared from two isotopically distinct feedstocks, and drillings from various ingot sections were dissolved in a mixture of HCl, HNO_3, and HF. Uranium isotope ratios in each solution were determined by inductively coupled plasma-mass spectrometry (ICP-MS); total dissolved solid content was 20 μg/L for determination of $^{238}U/^{235}U$, and 2 mg/L for determination of $^{236}U/^{235}U$ and $^{234}U/^{235}U$. Between-sample coefficients of variation (n=35) for $^{238}U/^{235}U$, $^{236}U/^{235}U$, and $^{234}U/^{235}U$ were 0.70%, 0.90%, and 1.4%, respectively. Linear regression indicated that $^{236}U/^{235}U$ *versus* $^{238}U/^{235}U$ and $^{234}U/^{235}U$ *versus* $^{238}U/^{235}U$ were significantly correlated at the 90% confidence interval; $^{234}U/^{235}U$ *versus* $^{236}U/^{235}U$ was significantly correlated at the 98% confidence interval. These data show that the samples were not isotopically homogeneous, and demonstrate that ICP-MS can be used to assess isotopic homogeneity, even in cases where aggregate variations in isotopic composition are on the order of 1%.

KEYWORDS: uranium, isotopic composition, ICP mass spectrometry

---

The world inventory of enriched uranium (EU, $^{235}U > 0.7\%$) is presently estimated at *ca.* 63,000 metric tons [*1*]. Much of this material (*ca.* 99%) is low enrichment uranium, which is used primarily as nuclear power reactor fuel. The remaining EU inventory, which consists of high enrichment uranium (HEU, $^{235}U > 20\%$), is an attractive target for illicit diversion and subsequent use in nuclear devices.

Isotopic "blending," i.e., dilution of HEU with natural or depleted uranium ($^{235}U \leq 0.7\%$), is being implemented as a means of minimizing the HEU inventory, thereby reducing the nuclear proliferation risk [2]. The blending process may involve uranium metals mixed at high temperature, uranyl nitrates mixed in solution, or uranium hexafluorides mixed in the gas phase; however, the mode of blending is irrelevant, so long as the final product is unattractive to potential nuclear proliferants.

---

[1] Chemist, Analytical Chemistry Laboratory/Chemical Technology Division, Argonne National Laboratory, 9700 South Cass Avenue, Argonne, IL 60439. Present Address: Union Carbide Corporation, P.O. Box 8361, South Charleston, WV 25303.

Blended uranium would be most unattractive when the final product is isotopically homogeneous. Homogeneity certification using conventional measurement technology, i.e., thermal ionization mass spectrometry (TIMS), though thorough, would also be time-consuming, and possibly expensive. As an alternative, one could use inductively coupled plasma-mass spectrometry (ICP-MS) in place of TIMS, sacrificing significant figures in the interest of speed, but this would work only if the acceptable "percent heterogeneity" falls within the capabilities of ICP-MS.

In this work, quadrupole ICP-MS was used for the determination of uranium isotopic composition in blended uranium metal. Between-sample variations in isotopic composition were on the order of 1%, but careful statistical analysis of the data revealed subtle isotopic differences between samples. These results show that ICP-MS can be successfully used in support of blended uranium production and certification.

Procedure

Material Preparation

Two isotopically distinct depleted uranium (DU) feedstocks were thoroughly blended at high temperature, and the resultant uranium metal ingot (20 kg) was sectioned. Drillings from each section were collected for isotopic analysis.

Sample Preparation

The uranium drillings were cleaned with HPLC grade methanol and dried. Then, the samples (10 to 100 mg each) were weighed and quantitatively transferred to PTFE beakers, where 5 mL of reagent water (18 MΩ-cm resistivity) and 2 mL of concentrated, high-purity hydrochloric acid (HCl) were added to each sample. The acid mixtures were heated gently on a hot plate for 30 to 45 min, and after this initial dissolution step, 2 mL of concentrated, high-purity nitric acid (HNO_3) were added to each sample. Gentle heating was continued until the solutions clarified (30 to 45 min).

After clarification, 4 mL of reagent water and 0.2 mL of concentrated, reagent grade hydrofluoric acid (HF) were added to each sample solution. The solutions were digested for 30 to 40 min on low heat to dissolve zirconium (an alloying element present in the feedstocks), and after cooling, the solutions were serially diluted with 1 M HNO_3 to obtain a total dissolved solid (TDS) concentration of 100 mg/L. Aliquots of these solutions were drawn and diluted 50-fold or 5000-fold with 1 M HNO_3 to facilitate determination of minor ($^{234-236}U$) and major ($^{235,238}U$) isotopes, respectively.

Mass Spectrometry

Uranium isotope ratios (*versus* ^{235}U) were determined over a period of one calendar week using a Plasmaquad II+ (VG Elemental, Winsford, UK) ICP mass spectrometer. Samples were introduced into the plasma ion source using a V-groove pneumatic nebulizer and a water-cooled double pass spray chamber. Typical operating conditions and general protocols for isotopic analysis with this instrument have been published elsewhere [*3*].

Isotope (atom) ratios were calculated from five consecutive 30 s signal integrations per sample. Signals were integrated in the "peak jumping" mode; dwell times per sweep were: 60 ms each for $^{234,236}U^+$, 30 ms for $^{235}U^+$, and 6 ms for $^{238}U^+$. The isotope ratio response of the mass spectrometer was calibrated with a 1M HNO_3 solution containing certified reference material (CRM) U010 (*ca.* 1 atom % ^{235}U, New Brunswick Laboratory, Argonne, IL). The calibrant solution was prepared such that its uranium concentration was identical to the TDS concentration of the sample solutions, and it was analyzed at the beginning, end, and at intervals within each sample group. Isotope ratio bias factors were calculated from the quotient of the actual and observed isotope ratios, and linear interpolation was used to compensate for within-day bias factor variations. The observed isotope ratios in the samples were multiplied by the interpolated bias factors to obtain corrected ratios.

On the first day of this study, the 20 μg/L TDS solutions (group 1) were analyzed to determine $^{238}U/^{235}U$. The 2 mg/L TDS solutions (group 2) were analyzed on the following day to determine $^{236}U/^{235}U$ and $^{234}U/^{235}U$. The calibrant solution and several arbitrarily selected samples from within each group were analyzed in triplicate to assess the accuracy and within-day stability of the isotope ratio determinations. On the two succeeding (non-consecutive) days of the experiment, the samples that had been analyzed in triplicate were re-analyzed to provide between-day (external) precision data.

Results

Isotope Ratio Precision and Accuracy

TABLE 1 -- *Isotope ratios for CRM U010.*

	$^{238}U/^{235}U$	$^{236}U/^{235}U$	$^{234}U/^{235}U$
Days 1 & 2	98.79	6.69×10^{-3}	5.54×10^{-3}
	103.9 [a]	6.71×10^{-3}	5.45×10^{-3}
	98.47	6.74×10^{-3}	5.43×10^{-3}
	98.22	6.76×10^{-3}	5.93×10^{-3} [a]
Day 3	98.68	6.77×10^{-3}	5.39×10^{-3}
Day 4	97.57	6.81×10^{-3}	5.39×10^{-3}
mean [b]	98.25	6.767×10^{-3}	5.419×10^{-3}
st. dev. [b]	0.59	4.1×10^{-5}	4.6×10^{-5}
certified	98.62	6.785×10^{-3}	5.390×10^{-3}

[a] outlier, rejected by Q test with 95% confidence.
[b] Day 1 & 2 data were averaged, then the between-day mean and standard deviation (shown) were calculated.

Using the Q test [*4*], outliers were rejected with 95% confidence from the within- and between-day isotopic analyses (Tables 1 through 3). Subsequent statistical analysis indicated that, in the case of CRM U010, the mean isotope ratios (within- and between-day) were indistinguishable from their known values at the 95% confidence interval. It was also found that differences among between-day and within-day isotope ratio variations were insignificant at the 95% confidence interval, except in the case of $^{238}U/^{235}U$ for sample 97-154-32. While these results do not necessarily prove that the sample isotope ratios were accurately determined, they do indicate that the calibration protocol compensated for within- and between-day variations in isotope ratio bias.

TABLE 2 -- *Isotope ratios (between day average) for selected samples.*

Sample ID	$^{238}U/^{235}U$	$^{236}U/^{235}U$	$^{234}U/^{235}U$
97-154-5	386.5 ± 3.9 [a]	(9.213 ± 0.043) ×10^{-3}	(4.744 ± 0.047) ×10^{-3}
97-154-14	386.6 ± 1.9	(9.193 ± 0.047) ×10^{-3}	(4.729 ± 0.083) ×10^{-3}
97-154-23	385.12 ± 0.64	(9.255 ± 0.029) ×10^{-3}	(4.724 ± 0.047) ×10^{-3}
97-154-32	386.1 ± 3.6	(9.17 ± 0.16) ×10^{-3}	(4.749 ± 0.023) ×10^{-3}

[a] mean ± standard deviation (n=3) is shown.

TABLE 3 -- *Isotope ratios (within day average) for selected samples.*

Sample ID	$^{238}U/^{235}U$	$^{236}U/^{235}U$	$^{234}U/^{235}U$
97-154-5	390.6 ± 1.1 [a]	(9.210 ± 0.031) ×10^{-3}	(4.727 ± 0.039) ×10^{-3}
97-154-14	387.8 ± 1.5	(9.211 ± 0.031) ×10^{-3}	(4.699 ± 0.028) ×10^{-3}
97-154-23	384.78 ± 0.33	(9.260 ± 0.072) ×10^{-3}	(4.676 ± 0.024) ×10^{-3}
97-154-32	389.36 ± 0.38	(8.98 ± 0.11) ×10^{-3}	(4.73 ± 0.10) ×10^{-3}

[a] mean ± standard deviation (n=3) is shown.

Isotopic Homogeneity

Using the Q test at 95% confidence, isotope ratios from one sample (out of 36) were rejected as outliers due to an excessively high $^{234}U/^{235}U$ result. For the remaining 35 samples, the aggregate (between-sample) mean for $^{238}U/^{235}U$ was 387.5, $^{236}U/^{235}U$ was 9.178 ×10^{-3}, and $^{234}U/^{235}U$ was 4.75 ×10^{-3}. The corresponding between-sample coefficients of variation (CVs) were 0.70%, 0.90%, and 1.4%, respectively. The between-sample variations were similar in magnitude to the within-sample variations (Table 3); however, one-way analysis of variance (ANOVA) [*4*] showed that, for $^{238}U/^{235}U$ ($F = 21$) and $^{236}U/^{235}U$ ($F = 9.7$), within- and between-sample variance was significantly different at the 95% confidence interval. These results, though derived from a small data set, suggest that the samples may have been isotopically different.

Subsequent linear least-squares regression of $^{236}U/^{235}U$ and $^{234}U/^{235}U$ *versus* $^{238}U/^{235}U$ indicated that correlations between the minor and major isotope ratios, though weak, were significant at the 90% confidence interval. Weak correlation was expected in this case, given the order-of-magnitude differences between $^{238}U/^{235}U$ and the other ratios. Regression of $^{234}U/^{235}U$ *versus* $^{236}U/^{235}U$ (Fig. 1) showed stronger correlation, significant at the 98% confidence interval. Population density plots and k-means cluster analysis of the data in Fig. 1 also identified bimodality in the $^{236}U/^{235}U$ distribution, which would be expected from incomplete mixing of the DU feedstocks. This observation, taken in combination with the ANOVA and linear regression data, strongly suggest that the samples were not of identical isotopic composition, despite the percent level between-sample variations.

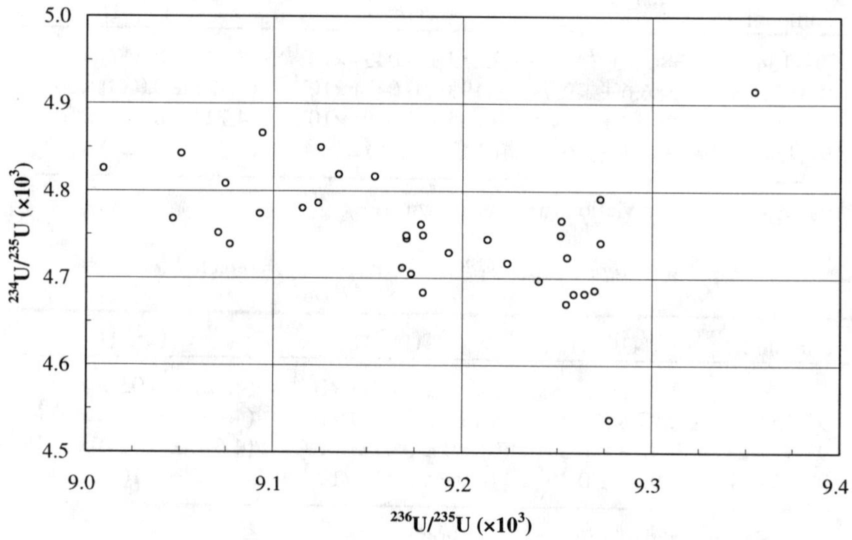

FIG. 1 -- *Correlation of* $^{234}U/^{235}U$ *with* $^{236}U/^{235}U$.

Conclusion

The data presented herein indicate that, with rigorous analysis of isotope ratio data, ICP-MS is capable of detecting significant differences in the isotopic composition of uranium samples, even in cases where aggregate (between-sample) coefficients of variation are on the order of 1%. The "limit of detection" for these differences was not quantified in this study; however, linear regression of minor isotope correlation plots (e.g., Fig. 1) appears to be a facile and powerful means of detecting such differences.

In the absence of a carefully designed sampling plan, no attempt was made to connect the between-sample isotopic variations with significant heterogeneities in the original uranium ingot. Moreover, even if we assumed that the between-sample variations were indicative of bulk heterogeneity, it remains to be seen if the level of heterogeneity would be operationally significant, i.e., an impediment to the production of blended uranium materials. It is hoped that, as blending operations mature and become more prevalent, these issues will be addressed by other analysts.

Acknowledgments

The author wishes to thank Paul Roach, Donald Graczyk, Doris Huff, Jane Marr, and Dennis Brady for helpful discussions and experimental support. This work was performed at Argonne National Laboratory, which is operated by the University of Chicago for the United States Department of Energy under contract number W-31-109-ENG-38.

Disclaimer

Reference to any specific commercial product, process, or service by trade name, trademark, manufacturer, or otherwise does not necessarily constitute or imply its endorsement, recommendation, or favoring by the U.S. Government or any agency thereof. The views and opinions of document authors do not necessarily state or reflect those of the U.S. Government or any agency thereof, Argonne National Laboratory, or the University of Chicago.

References

[*1*] 1996 Annual Report, International Atomic Energy Agency, Vienna, 1997, p. 78.

[*2*] U.S. Department of Energy, "Record of Decision for the Disposition of Surplus Highly Enriched Uranium Final Environmental Impact Statement," *Federal Register*, Vol. 61, No. 151, 1996, pp. 40619-40629.

[*3*] Crain, J.S. and Alvarado, J., *Journal of Analytical Atomic Spectrometry*, Vol. 9, 1994, pp. 1223-1227.

[*4*] Miller, J.C. and Miller, J.N., *Statistics for Analytical Chemistry, 3rd ed.*, Ellis Horwood PTR Prentice Hall, New York, 1993, chap. 3.

Ian Bowen,[1] Andrew J. Walder,[1] Tom Hodgson[2] and R. R. Parrish[3]

HIGH PRECISION AND HIGH ACCURACY ISOTOPIC MEASUREMENT OF URANIUM USING LEAD AND THORIUM CALIBRATION SOLUTIONS BY INDUCTIVELY COUPLED PLASMA-MULTIPLE COLLECTOR-MASS SPECTROMETRY

REFERENCE: Bowen, I., Walder, A. J., Hodgson, T., and Parrish, R. R, **"High Precision and High Accuracy Isotopic Measurement of Uranium Using Lead and Thorium Calibration Solutions by Inductively Coupled Plasma-Multiple Collector-Mass Spectrometry,"** *Applications of Inductively Coupled Plasma-Mass Spectrometry to Radionuclide Determinations: Second Volume, ASTM STP1344,* R. W. Morrow and J. S. Crain, Eds., American Society for Testing and materials, 1998.

ABSTRACT: A novel method for the high accuracy and high precision measurement of uranium isotopic composition by Inductively Coupled Plasma-Multiple Collector-Mass Spectrometry is discussed. Uranium isotopic samples are spiked with either thorium or lead for use as internal calibration reference materials. This method eliminates the necessity to periodically measure uranium standards to correct for changing mass bias when samples are measured over long time periods. This technique has generated among the highest levels of analytical precision on both the major and minor isotopes of uranium. Sample throughput has also been demonstrated to exceed Thermal Ionization Mass Spectrometry by a factor of four to five.

KEYWORDS: isotope ratio measurement, uranium, nuclear fuel enrichment, ICP-MS, ICP-MC-MS.

The production of uranium fuel for civil nuclear reactors involves enrichment of the fissile ^{235}U isotope. Such an enrichment process is the core business activity of URENCO (Capenhurst) Ltd. Uranium in the form of gaseous uranium hexafluoride (UF_6) is fed into gas centrifuge plants to enrich the ^{235}U isotope. It is important to the economic viability of the enrichment process to control the abundance of the ^{235}U in the product within the

[1]Product manager MC-ICP-MS and Technical director, VG Elemental, Ion Path, Road Three, Winsford, Cheshire, CW7 3BX, UK.

[2]Technical specialist, URENCO (Capenhurst) Ltd, Mass Spectrometry Lab., Building 202, Capenhurst Works, Capenhurst, Chester, CH1 6ER , UK.

[3]Professor, Department of Geology, University of Leicester and NERC Isotope Geosciences Laboratory, British Geological Survey, Keyworth, Notts, NG12 5GG, UK.

customer's specification. This imposes an exacting tolerance for the ^{235}U abundance. The mass spectrometer and associated measurement techniques described in this paper greatly enhance the efficiency of such ^{235}U abundance measurements.

In addition to specifying the abundance of the ^{235}U isotope, customers also limit the quantities of the minor isotopes of uranium (^{232}U, ^{233}U, ^{234}U and ^{236}U) in the fuel. To ensure that a provider of enrichment is not restricted in the choice of feed stock, it is vital that the abundance of the minor isotopes (particularly ^{234}U and ^{236}U) are accurately known. Mass spectrometry techniques are used to measure the abundance of the ^{234}U, ^{235}U and ^{236}U isotopes at various stages of the enrichment process and also in the final product container.

Isotopic measurements within the nuclear industry are traditionally performed by Thermal Ionization Mass Spectrometry (TIMS). This type of instrumentation is generally considered as the benchmark technique for accurate and precise isotopic ratio measurements. TIMS is, however, complicated by tedious and time consuming sample preparation. Uranium isotopic measurements are also complicated by time dependent sample fractionation effects, this phenomenon places a limit on the accuracy and precision that can be achieved.

The measurement of isotope ratios by Inductively Coupled Plasma Mass Spectrometry (ICP-MS) offers significant advantages over TIMS. These include much reduced sample preparation, simplified sample chemistry, faster sample throughput and the absence of any time dependent sample fractionation effects. Conventional ICP-MS instruments equipped with a single ion detector are in abundant use for isotopic ratio measurements, however, their use has been severely limited by the accuracy and precision limitations associated with the sequential measurement of isotopic composition. The introduction of ICP-Multiple Collector-Mass Spectrometers (ICP-MC-MS) has for the first time allowed TIMS levels of isotopic precision with all the advantages of an ICP ion source. A summary of the instrumentation and application details are summarised in a review chapter [*1*].

While most isobaric and molecular interferences are not an issue for uranium hexaflouride production, there will be other industrial processes where interferences exist. Oxide and hydride molecular interferences can be reduced and in most cases eliminated by desolvation sample introduction techniques. ICP-MC-MS can correct for isobaric interferences of stable isotopes such as ^{144}Sm on ^{144}Nd, ^{176}Lu on ^{176}Hf and ^{204}Hg on ^{204}Pb [*2*]. However, correction of radiogenic species is not possible and would require the chemical removal of the interference prior to analysis.

A VG Elemental ICP-MC-MS (The VG Plasma 54) has been customised specifically for the nuclear industry. Its isotope ratio performance has been evaluated by URENCO (Capenhurst) Ltd for the analysis of both the major and minor isotopes of uranium. Sample throughput, automation, sample preparation and cost of analysis were also investigated.

The measurement of uranium isotopic ratios have previously been reported by ICP-MC-MS [*3*]. Throughout this study mass bias was determined with the measurement of a known uranium reference material. This value was then used for the correction of subsequent samples. This is a technique routinely utilized in TIMS and is referred to as an external mass fractionation correction procedure. However, the use of an external calibration for mass bias was found to limit the precision and accuracy of the uranium isotopic measurement, especially when a single external calibration result was used over a long time period.

A study by Taylor et al [*4*] utilized IRMM 072 series of uranium reference solutions to determine the most appropriate law to correct for mass bias. Both the power and exponential laws were found to produce accurate corrections for mass bias. For our study a power law was used throughout.

Internal mass bias correction is routinely used in isotopic studies both by TIMS and ICP-MC-MS. For example, the ^{144}Nd and ^{146}Nd isotopes of naturally occurring neodymium are stable and therefore the isotopic ratio is fixed in nature, a calculation of mass bias is made by comparing the expected with measured value. This comparison is used to simultaneously correct the other isotopic measurements of neodymium. Although the ^{235}U:^{238}U isotopic ratio is known for natural uranium, this methodology cannot be applied to the measurement of uranium in nuclear materials because of the inherent isotopic variation produced by the enrichment or fission process.

A novel method for determining high precision and high accuracy isotopic measurements of uranium was evaluated in this study. This technique exploits the unique feature of an ICP source that allows one element to be used as a mass bias correction for another element. For example, thallium has been used as an accurate correction for lead isotopic analysis by ICP-MC-MS [*5*]. While the ionization efficiencies of lead, thorium and uranium differ, mass bias variation of the isotopes of uranium will demonstrate a similar characteristic as lead and thorium. This study will describe how the isotopes of lead and thorium can be used as an internal correction for the measurement of uranium isotopes.

Experimental

Instrumentation

The ICP-MC-MS (VG Plasma 54) utilized in this study was a double-focusing magnetic sector instrument [*6, 7*]. It was equipped with a seven Faraday collector array and a high performance Daly ion counting multiplier system. The Faraday collector configuration consisted of four high mass; one axial and two low mass detectors.

The Faraday collectors low 2, low 1, high 1, high 2 and high 4 were used to measure the major isotopes. The high 3 collector was not required during any of the measurement sequences. The axial Faraday was constantly in the lowered position to allow the Daly multiplier system for the detection of the minor ^{234}U and ^{236}U isotopes.

Samples were introduced into the ICP source using a 44 position autosampler. Solution uptake from the autosampler test tubes was regulated with a peristaltic pump. This pump also extracted the excess solution from the spray chamber and supplied clean acid solution to the wash position of the autosampler. Autosampler control, solution uptake, and washout were fully automated.

Analytical Procedure

Three different hydrolysed uranium hexafluoride samples were selected for measurement. The isotopic composition of these solutions were representative of the range of samples encountered in uranium nuclear fuel enrichment facilities. The major and minor isotopic abundances of these samples were well characterised. The samples were referenced internally as BX294, BX214/11 and REIMEP 89 and will henceforth be referenced as samples A, B and C. An in house reference material was also used within this study and was known as BX264/3. The atom percents of all the uranium samples used are detailed (Table 1).

TABLE 1 - *Atom Percents of the Uranium Samples*

Sample	^{234}U	^{235}U	^{236}U
BX 264/3	0.0414 ± 0.0004	3.5390 ± 0.0018	0.0101 ± 0.0001
A (BX 294)	0.0726 ± 0.0007	5.7994 ± 0.0029	0.0135 ± 0.0001
B (BX 214)	0.00548 ± 0.00006	0.72025 ± 0.00022	<0.00005
C (REIMEP)[1]	0.0020	0.3194	0.0130

[1]Uncertified material (quoted abundance values only).

Each uranium sample was diluted to a concentration of 2 μg mL^{-1} of uranium using de-ionised water. This concentration produced a total measured uranium ion current of $6x10^{-11}$A with a solution uptake rate of 0.4 mL min^{-1}. No high efficiency sample introduction systems were evaluated as this particular application is not sample limited.

The measurement sequence was designed to determine the internal precision of individual sample measurements, the external precision of multiple sample measurements, to test the memory characteristics of the ICP source and to determine the speed of sample throughput. After a single measurement of the in-house reference material, ten aliquots of each of the three samples were measured in a A; B; C; A; B; C... cyclic sequence.

Each measurement block (Tables 2 and 3) generally consisted of ten measurements of five second integration periods. With the sample uptake, peak centering, baseline measurements and subsequent block measurements, the time for each individual sample analysis was 8 to 9 minutes.

The speed of the peristaltic pump controlling the sample uptake was automatically varied at different stages. A slow speed of 12 rpm was used for the isotope ratio measurement of each sample. The pump was accelerated to 48 rpm for the wash cycle between samples. A 10% HNO_3 solution was used for the wash and a five minute washout period between samples was sufficient to reduce the presence of the previous sample by three orders of magnitude. The amount of uranium sample and acid wash solutions consumed per measurement were approximately 5 mL and 10 mL respectively.

Uranium Isotopic Analysis Using a Thorium Calibration

We take a new approach to the calibration of uranium isotopic measurement and related mass fractionation. This approach uses a two isotope (^{230}Th:^{232}Th) solution to monitor the mass fractionation correction. Artificially produced thorium is available and its use as an internal calibration material for the determination of uranium isotopic composition was investigated here. The isotopic ratio of the thorium used was approximately ^{230}Th:^{232}Th = 0.01 at a concentration of 2 µg mL^{-1}. Thorium was added to the uranium reference solution and the three uranium samples prior to the division into the ten aliquots. The Faraday collectors were positioned to simultaneously measure the uranium and thorium isotopes (Table 2). A two stage peak jumping routine was used to allow measurement of the minor ^{234}U and ^{236}U isotopes on the Daly multiplier detector.

TABLE 2 - *Measurement sequence for uranium using thorium as a calibration material*

Collector	Low 2	Low 1	Axial Daly	High 1	High 2	High 3	High 4
Block 1	^{230}Th	^{232}Th	^{234}U	^{235}U			^{238}U
Block 2			^{236}U		^{238}U		

The absolute isotopic composition of the thorium was not determined in an absolute sense and is not important for the methodology followed. Instead, a uranium reference material (BX264/3) at 2 µg mL^{-1} was spiked with a similar concentration of the thorium solution. Comparison of the measured ^{235}U:^{238}U value with the reference value was used to determine the magnitude of mass bias. This value of mass bias was then used to correct the measured ^{230}Th:^{232}Th. This corrected thorium value was then determined and used in subsequent analysis as the internal reference.

The uranium/thorium samples A, B and C were then measured. The measured ^{230}Th:^{232}Th value for each sample was compared with the corrected thorium value determined from the uranium reference material measurement. A new mass bias value was thus determined. This mass bias value could then be used to internally correct the measured isotope ratios of the uranium samples.

Uranium Isotopic Analysis Using a Lead Calibration

As an alternative calibration methodology a 2 µg mL^{-1} solution of a JMC lead reference material was added to all of the uranium samples. Again, the lead was added to the three uranium samples prior to the division into the ten aliquots. The isotopic composition of the lead used was approximately ^{207}Pb:^{208}Pb = 0.4. Again, the absolute isotopic composition of the lead was not known and is not important for the methodology followed. A uranium reference material (BX264/3 at 2 µg mL^{-1}) was spiked with a similar concentration of the lead solution. The Faraday collectors were positioned to measure all the uranium and ^{207}Pb and ^{208}Pb isotopes in the configuration shown (Table 3). As the mass separation between lead and uranium is too large to allow a simultaneous measurement a third peak jumping routine was used. This required an initial measurement of lead and subsequent measurements of uranium. A comparison of the measured ^{235}U:^{238}U value with the reference value was used to determine the magnitude of mass bias. This value of mass bias was then used to correct the measured ^{207}Pb:^{208}Pb. This corrected lead value was then stored for subsequent use.

The uranium/lead samples A, B and C were then measured. The measured ^{207}Pb:^{208}Pb value for each sample was compared with the corrected lead value determined from the uranium reference material. A new mass bias value was thus determined. This mass bias value was then used to correct the isotopes of uranium within the sample.

TABLE 3 - *Measurement sequence for uranium using a lead calibration*

Collector	Low 2	Low 1	Axial Daly	High 1	High 2	High 3	High 4
Block 1	^{207}Pb	^{208}Pb					
Block 2			^{234}U	^{235}U			^{238}U
Block 3			^{236}U		^{238}U		

Results and Discussion

The measurement results of the isotopic composition of samples A, B and C using the thorium calibration technique are shown (Table 4). Similarly, the results for the lead calibration technique are shown (Table 5). The values in parenthesis along side each result indicates the percent standard error (%SE 1σ) for each sample aliquot measurement. The mean values along with the percent relative standard deviation (%RSD) for each atom percent are shown at the bottom of each sample type. The levels of analytical precision achieved are superior to that demonstrated by TIMS. The calibration technique utilized produces accurate and precise data for both the major and minor isotopes of uranium.

The two methods of internal mass bias correction demonstrated similar accuracy and precision. The corrected values of sample A give ^{235}U atom percents of 5.7983 ± 0.0015 and 5.8004 ± 0.0019, which are in agreement of the accepted value of 5.7994 ± 0.0029. The corrected values of sample B give ^{235}U atom percents of 0.72030 ± 0.00026 and 0.72043 ± 0.00027, which are in agreement of the accepted value of 0.72025 ± 0.00022. The corrected values of sample C give ^{235}U atom percents of 0.31974 ± 0.00019 and 0.31959 ± 0.00021, which are in agreement of the accepted value of 0.3194. For the uranium/thorium samples, the total time for the measurement of the internal reference material followed with the measurement of the 30 samples (10 aliquots of each of the 3 samples) was approximately 7 hours, whereas for the uranium/lead samples the measurement time was approximately 9 hours. This represents a factor of four to five increase in sample throughput compared to thermal ionization techniques.
The uranium/thorium experiment was repeated on subsequent days to confirm its reproducibility. A summary of three days results are shown (Table 6).

The ^{234}U and ^{236}U results agree with the reference values to within 1.5% demonstrating accuracy, this is partly due to the good abundance sensitivity of the instrument and linearity of the multiplier system. The abundance sensitivity was measured previously by measuring the contribution of the ^{238}U at mass 237 using the Daly multiplier system. The abundance sensitivity was found to be approximately 1 ppm at the normal working vacuum in the instrument analyzer of 4 x 10^{-10} kPa.

The method can also be used to calibrate the absolute isotopic composition of the synthetic dual isotope thorium solution. This can be done by a precise analysis of the thorium solution spiked with uranium and lead isotopic standards whose values are independently known. Comparison of measured to known uranium and lead isotopic compositions will yield the mass dependence of isotopic fractionation in the region of thorium by interpolation. The absolute isotopic composition of thorium can then be

TABLE 4 - *Atom percents of the thorium/uranium samples*

Sample A	Atom % 234U		Atom % 235U		Atom % 236U	
1	0.07249	(0.06)	5.79711	(0.006)	0.01348	(0.03)
2	0.07271	(0.03)	5.79896	(0.007)	0.01346	(0.04)
3	0.07274	(0.04)	5.79827	(0.005)	0.01346	(0.03)
4	0.07238	(0.05)	5.79866	(0.004)	0.01347	(0.04)
5	0.07264	(0.04)	5.79876	(0.007)	0.01348	(0.04)
6	0.07245	(0.03)	5.79835	(0.008)	0.01348	(0.06)
7	0.07270	(0.05)	5.79969	(0.008)	0.01348	(0.06)
8	0.07238	(0.06)	5.79748	(0.006)	0.01352	(0.04)
9	0.07273	(0.04)	5.79800	(0.007)	0.01347	(0.04)
10	0.07259	(0.03)	5.79799	(0.007)	0.01348	(0.05)
Mean	0.07258		5.79833		0.01348	
% RSD	0.20		0.013		0.11	
Sample B	**Atom % 234U**		**Atom % 235U**		**Atom % 236U**	
1	0.00540	(0.05)	0.72046	(0.011)	0.00000	
2	0.00539	(0.06)	0.72035	(0.010)	0.00000	
3	0.00540	(0.08)	0.72016	(0.009)	0.00000	
4	0.00541	(0.08)	0.72021	(0.010)	0.00000	
5	0.00540	(0.05)	0.72049	(0.010)	0.00000	
6	0.00540	(0.07)	0.72028	(0.010)	0.00000	
7	0.00539	(0.07)	0.72028	(0.009)	0.00000	
8	0.00538	(0.06)	0.72014	(0.009)	0.00001	
9	0.00543	(0.06)	0.72021	(0.010)	0.00002	
10	0.00541	(0.09)	0.72042	(0.010)	0.00000	
Mean	0.00540		0.72030		0.00000	
% RSD	0.26		0.018			
Sample C	**Atom % 234U**		**Atom % 235U**		**Atom % 236U**	
1	0.00188	(0.10)	0.31980	(0.009)	0.01296	(0.04)
2	0.00188	(0.07)	0.31974	(0.010)	0.01295	(0.04)
3	0.00188	(0.12)	0.31992	(0.010)	0.01296	(0.05)
4	0.00188	(0.12)	0.31971	(0.011)	0.01296	(0.04)
5	0.00188	(0.14)	0.31975	(0.012)	0.01296	(0.04)
6	0.00188	(0.09)	0.31956	(0.009)	0.01298	(0.04)
7	0.00187	(0.12)	0.31974	(0.012)	0.01297	(0.05)
8	0.00188	(0.08)	0.31965	(0.013)	0.01294	(0.05)
9	0.00187	(0.14)	0.31975	(0.009)	0.01294	(0.04)
10	0.00188	(0.19)	0.31980	(0.012)	0.01295	(0.04)
Mean	0.00188		0.31974		0.01296	
% RSD	0.23		0.030		0.09	

Values in parentheses are %SE

calculated and the measured thorium ratio corrected. This approach could be used to produce a thorium isotopic reference material to monitor fractionation in thermal ionization mass spectrometry, a technique used extensively in age determinations in the earth sciences.

TABLE 5 - *Atom percents of the uranium/lead samples*

Sample A	Atom % 234U		Atom % 235U		Atom % 236U	
1	0.07237	(0.06)	5.79952	(0.005)	0.01343	(0.06)
2	0.07234	(0.03)	5.80071	(0.004)	0.01343	(0.07)
3	0.07239	(0.04)	5.80042	(0.003)	0.01343	(0.09)
4	0.07234	(0.05)	5.80101	(0.004)	0.01343	(0.07)
5	0.07222	(0.03)	5.80008	(0.006)	0.01338	(0.06)
6	0.07228	(0.05)	5.79823	(0.005)	0.01340	(0.05)
7	0.07213	(0.06)	5.80039	(0.004)	0.01341	(0.07)
8	0.07209	(0.04)	5.80178	(0.003)	0.01340	(0.06)
9	0.07217	(0.04)	5.80073	(0.003)	0.01339	(0.06)
10	0.07228	(0.04)	5.80092	(0.004)	0.01341	(0.05)
Mean	0.07226		5.80038		0.01341	
% RSD	0.15		0.017		0.14	

Sample B	Atom % 234U		Atom % 235U		Atom % 236U
1	0.00541	(0.05)	0.72031	(0.007)	0.00000
2	0.00540	(0.06)	0.72065	(0.006)	0.00000
3	0.00540	(0.08)	0.72039	(0.007)	0.00000
4	0.00541	(0.05)	0.72036	(0.005)	0.00000
5	0.00539	(0.07)	0.72024	(0.007)	0.00000
6	0.00540	(0.06)	0.72051	(0.006)	0.00000
7	0.00540	(0.05)	0.72061	(0.008)	0.00000
8	0.00540	(0.08)	0.72030	(0.007)	0.00000
9	0.00540	(0.06)	0.72053	(0.007)	0.00000
10	0.00541	(0.09)	0.72041	(0.009)	0.00000
Mean	0.00540		0.72043		0.00000
% RSD	0.13		0.019		

Sample C	Atom % 234U		Atom % 235U		Atom % 236U	
1	0.00189	(0.10)	0.31975	(0.008)	0.01291	(0.05)
2	0.00188	(0.09)	0.31940	(0.009)	0.01292	(0.05)
3	0.00188	(0.11)	0.31965	(0.008)	0.01292	(0.06)
4	0.00188	(0.12)	0.31953	(0.012)	0.01290	(0.08)
5	0.00188	(0.08)	0.31953	(0.008)	0.01293	(0.04)
6	0.00188	(0.12)	0.31965	(0.011)	0.01293	(0.04)
7	0.00188	(0.14)	0.31956	(0.009)	0.01292	(0.06)
8	0.00187	(0.09)	0.31958	(0.014)	0.01290	(0.05)
9	0.00189	(0.11)	0.31974	(0.010)	0.01292	(0.05)
10	0.00188	(0.12)	0.31953	(0.011)	0.01292	(0.06)
Mean	0.00188		0.31959		0.01292	
% RSD	0.26		0.033		0.09	

Values in parentheses are %SE

TABLE 6 - Three day trial of uranium/thorium samples

Sample	Atom %	Day 1 mean	% RSD	Day 2 mean	% RSD	Day 3 mean	% RSD
	234U	0.07258	0.020	0.07284	0.14	0.07287	0.23
A	235U	5.79833	0.013	5.79935	0.011	5.79993	0.014
	236U	0.01348	0.011	0.01353	0.16	0.01351	0.11
	234U	0.00540	0.26	0.00541	0.25	0.00541	0.33
B	235U	0.72030	0.018	0.72042	0.024	0.72027	0.028
	236U	0.00000		0.00000		0.00000	
	234U	0.00188	0.23	0.00188	0.38	0.00187	0.26
C	235U	0.31974	0.030	0.31975	0.021	0.31972	0.030
	236U	0.01296	0.09	0.01296	0.19	0.01293	0.13

Conclusion

A novel technique utilizing both thorium and lead as isotopic calibration materials for the isotopic analysis of uranium has been successfully demonstrated. The levels of external and internal precision on both the major and minor isotopes of uranium have been shown to be superior to that achieved by TIMS.

References

[1] Platzner, I. T., "*Modern Isotope Ratio Mass Spectrometry*", John Wiley & Sons, 1997, chapter 4, pp. 83-108.

[2] Walder, A. J., Platzner, I., Freedman, P. A., "Isotope Ratio of Lead, Neodymium and Neodymium-Samarium Mixtures, Hafnium and Hafnium-Lutetium Mixtures With a Double Focusing Multiple Collector Inductively Coupled Plasma Mass Spectrometer," *Journal of Analytical Atomic Spectrometry*, 1993, Vol. 8, pp. 19-23.

[3] Walder, A. J., and Hodgson, T., "The Isotopic Ratio Measurement of Uranium in the Form of Hydrolzed Uranium Hexafluoride by Inductively Coupled Plasma Multiple Collector Mass Spectrometry," *Applications of Inductively Coupled Plasma-Mass Spectrometry to Radionuclide Determinations, ASTM STP 1291*, R. W. Morrow and J. S. Crain, Eds., American Society for Testing and Materials, West Conshohocken, 1995, pp. 20-25.

[4] Taylor, P. D. P., de Bièvre, P., Walder, A. J., and Entwistle, A., *Journal of Analytical Atomic Spectrometry*, 1995, Vol. 10, pp. 395-398.

[5] Walder, A. J., Koller, D., Reed, N. M., Hutton, R. C., and Freedman, P. A., "Isotope Ratio Measurement by Inductively Coupled Plasma Multiple Collector Mass Spectrometer Incorporating a High Efficiency Nebulization System," *Journal of Analytical Atomic Spectrometry*, 1993, Vol. 8, pp. 1037-1041.

[6] Walder, A. J., and Freedman, P. A., "Isotopic Ratio Measurement Using a Double Focusing Magnetic Sector Analyser With an Inductively Coupled Plasma as an Ion Source," *Journal of Analytical Atomic Spectrometry*, 1992, Vol. 7, pp. 571-575.

[7] Haliday, A. N., Lee D.-C., Christensen, J. N., Walder, A. J., Freedman, P. A., Jones C. E., Hall C. M., Yi, W., and Teagle, D. "Recent Developments in Inductively Coupled Plasma Magnetic Sector Multiple Collector Mass Spectrometry," *International Journal of Mass Spectrometry and Ion Processes*, 1995, 146/147, pp. 21-33

Laura L. Tovo[1], Joseph W. Clymire[1], William T. Boyce[1], and W. Frank Kinard[2]

ACTINIDE, ELEMENTAL, AND FISSION PRODUCT MEASUREMENTS BY ICPMS AT THE SAVANNAH RIVER SITE

REFERENCE: Tovo, L.L., Clymire, J.W., Boyce, W.T. , and Kinard, W.F., **"Actinide, Elemental, and Fission Product Measurements by ICPMS at the Savannah River Site,"** *Applications of Inductively Coupled Plasma-Mass Spectrometry to Radionuclide Determinations: Second Volume, ASTM STP 1344*, R.W. Morrow and J. S. Crain, Eds., American Society for Testing and Materials, 1998.

ABSTRACT: A VG Elemental Inductively coupled plasma-mass spectrometer (ICPMS), PlasmaQuad 1 (PQ1) Model No. 4, installed in a radiological controlled hood, is used by the Savannah River Technology Center to provide non-routine mass measurements for environmental monitoring, waste tank characterization studies, isotope ratios for criticality determinations, and the measurement of elemental, fission product, and actinide mass distributions of the glass product from the Defense Waste Processing Facility (DWPF).

Modifications to improve instrument reliability, sample preparation, and data handling, as well as modifications to the laboratory that permit measurements in a radioactive environment, will be discussed. Based on our operating experience, two laboratory facilities are being prepared for additional instruments to operate in a radioactive environment. A separate instrument is being installed for non-radioactive measurements and method development.

KEYWORDS: ICP-MS, radiochemical methods, actinides, fission products, high level radioactive waste

[1] Westinghouse Savannah River Company, Building 773-A, Aiken, SC 29802.
[2] Department of Chemistry, College of Charleston, Charleston, SC 29424.

Inductively coupled plasma mass spectrometry (ICPMS) is used across the Savannah River Site (SRS) to provide information for many site missions such as waste processing, environmental remediation, and reactor basin stabilization. The procedures developed over the last seven years provide site customers with relatively fast and reliable analyses to satisfy their process needs. The Savannah River Technology Center (the research and development laboratory for the site) currently has three instruments, a VG Elemental first generation PQ1 #4 and 2 VG Elemental PQ2+'s. The PQ1 and one of the PQ2's are housed in hoods and used for both radioactive and non-radioactive samples. The second PQ2 is used exclusively for research, procedure development and non-regulatory environmental samples.

The production environment at the SRS necessitates minimizing sample analysis time while maximizing quality from the analytical laboratory. SRS high level waste samples are typically high in salt content (~ 5 molar) and may contain large amounts of aluminum, iron, and organic species. These complex matrices complicate the radionuclide characterization of the sample by conventional counting methods [*1*] and therefore would normally require time-consuming sample pretreatment to eliminate matrix effects. However, in previous reported work [*1,2*], ICPMS was shown to provide rapid reliable analysis for actinides and fission products in the SRS high level wastes with minimal sample pretreatment. For the analysis of actinides and fission products, a Quickbasic™ [3] program was written to extract mass spectral data from the binary format of the ICPMS instrument data files. This data could then be imported to Grams/386™ [4] for visual examination or to an EXCEL™ [3] spreadsheet for quantitative analysis. The lack of readily available certified standards or known isotopic mixtures for all possible radioactive analytes posed a problem in the quantitative analysis of actinides and fission products. The analytical procedure developed to overcome this problem included a knowledge of nuclear properties, process information, and instrumental response. Common multielement standards were used to calibrate for natural abundance elements. The assumption was then made that the mass sensitivities over a small mass range are quite similar and therefore the mass bias is small over that range. The uncalibrated masses can then be estimated using the calibrated mass sensitivities. The application of this analysis method to SRS high level waste samples resulted in the determination of actinide and fission product concentrations with 25% precision at the 1 ppb level.[*2*]

Since commercial software did not exist for the analysis of non-natural and variable isotopic abundance elements, this first generation data reduction method provided a significant advantage in the analysis of SRS high level waste. Still, data handling required a high degree of human interaction with many manual manipulations in the EXCEL spreadsheet. Reduction in analysis time and the need for an audit trail of the calculations required a more user friendly and automated data reduction package.

In this paper, we present refinements to the first generation software that automates data reduction as well as hardware enhancements that improve instrument reliability and sensitivity. Experimental data is also presented to support the mass sensitivity assumption used in previous work [*1,2*]. The efficiency gained from these improvements proved beneficial in analyses performed for low activity and high level

---

[3] Trademark of Microsoft Corporation, One Microsoft Way, Redmond, WA 98052-6399.

[4] Trademark of Galactic Industries Corporation, 395 Main St., Salem, NH 03079.

waste processing treatability studies for the Hanford Tank Waste Remediation program. For this program, chemical, radionuclide, and physical property characterization was required on various Hanford waste streams. Like SRS high level waste, the Hanford waste streams are typically high in salt content, aluminum and organics and therefore the characterization is complicated by matrix effects. Published work [*6,7*] on the characterization of Hanford waste used ion chromatography coupled with ICPMS to eliminate matrix effects and interferences. Using the analysis procedure presented in this paper, the accuracy of a separation/ICPMS method is potentially sacrificed, but reliable and timely process information is obtained as well as the identification of of elements added during processing or present as contaminants.

Analytical Procedures and Instrumentation

Instrumentation

The containment hoods on the PQ1 and PQ2+ enclose the torch box, sampling cones, slide valve, peristaltic pump and optional autosampler. The sample effluent is collected in waste bottles or sent directly to the low level drain as required by the waste handling requirements of the sample. Electronics and vacuum pumps are located outside of the radioactive hood. Reported [*2,3*] and repeated in-house experience in vacuum system cleaning has demonstrated that the instrument pumps, detector region, quadrupole region and lens stack vacuum housing is free from detectable radioactive contamination. Fixed and smearable contamination is found on the cones. Within the lens stack, the photostop and first two lenses receive the majority of the contamination.

Routine sample introduction is accomplished with a Meinhard nebulizer, water cooled Scott spray chamber, and peristaltic pump operating at 1 mL per minute. Typically, uptake times of 120 seconds, 60 second acquisition times, and 0 rinse time is used for analysis of samples. The long uptake time is used to wash the previous sample from the introduction system. With the "dirty matrices" that we encounter, this method of sample introduction appears to work better than rinsing with 1% nitric acid or deionized water.

Detector lifetime is substantially increased by operating in the pulse counting mode only, and eliminating any response in excess of 10^6 in any single channel. As a rule of thumb, we measure only trace masses at or above mass 41. The instrument software has several measurement possibilities. Sample data may be collected in the scan mode, the peak jumping mode or the single ion monitoring mode. Both single ion monitoring and peak jumping require prior knowledge of the sample composition. The scan mode can detect all masses from mass 4 through mass 250, and is routinely used for unknown samples. Normally, a scan is performed at 1300 watts RF power. Data is acquired using the following parameters: 20 channels/amu , 320 microsecond dwell/channel, and a minimum of 3 scans per sample.

The PQ1 instrument operational reliability was significantly improved, from < 50% available to > 80% available, by upgrading the quadrupole rf generator and control electronics to equivalent PQ2+ components.

Dissolutions

Glass and sludge samples are dissolved using aqua regia, fusion, and microwave dissolutions. High level waste and radioactive glass samples are dissolved in the "high level caves" and diluted to within radioactivity control limits for analysis outside the remote facilities. Further dilution prior to ICPMS analysis is based on dissolved solids content, effects of matrix elements, and analyte concentration. As a rule, the analyte concentration is made to be >1 ppb but < 100 ppb. When necessary, matrix matching and spike recovery techniques from submitted samples are performed to verify analytical methodology. All samples are diluted with 1% nitric acid and spiked with 50 to 100 ppb of internal standard (indium, bismuth, scandium, or thallium). The internal standard is used to correct for variations in signal caused by environmental changes and matrix effects.

Results and Discussion

Multielement calibration solutions are a convenient way to calibrate the ICP-MS when a large number of elements or isotopes need to be determined simultaneously. In previous work [1,2], the assumption was made that reasonable estimates of sensitivities could be made for isotopes that are not present in calibration solutions. To support this assumption, the sensitivities of masses 130 through 180, 203 through 208, 232, 235, and 238 were determined independently. This mass region was chosen since the actinides and most of the high mass fission products fall within this range. Elemental concentration standards purchased from High Purity, Inc., Charleston, SC, were used to determine the isotopic sensitivities. Solutions of 1, 10, and 100 ppb of each element were prepared from the certified standard. Each solution was spiked with 50 ppb indium as an internal standard. With the instrument optimized for response, each standard solution was run using a full mass scan to assure the data collection timing was the same for each sample. The raw data was normalized to the indium internal standard. Using linear regression, the slopes of the concentration response curve were obtained for each isotope and are plotted in Figure 1.

The fission product region data in Figure 1 shows that over short mass ranges, the calibration sensitivities are approximately linear with mass. Thus, for isotopes not present in calibration solutions, sensitivities can be estimated from a linear regression of calibration sensitivities collected from neighboring isotopic concentration data. For example, Sm-151 is a fission product isotope that does not occur naturally and, therefore, would not be in the calibration standard. A linear regression of the calibrated Sm masses, 147 - 149 and 152 could be used to provide an estimate of the mass 151 sensitivity. The estimated mass 151 sensitivity would then be used to calculate the Sm 151 concentration. The sensitivities at masses 133, 138, and 139 in Figure 1 show larger deviations from linearity than the higher masses. The source of these deviations is not known at this time.

In the actinide region, the data in Figure 1 shows little or no variation of the sensitivity with mass. Therefore, the sensitivity of U-238 is used to estimate the uncalibrated actinide mass sensitivities. These estimated mass sensitivities are then used to calculate the concentrations of the actinides.

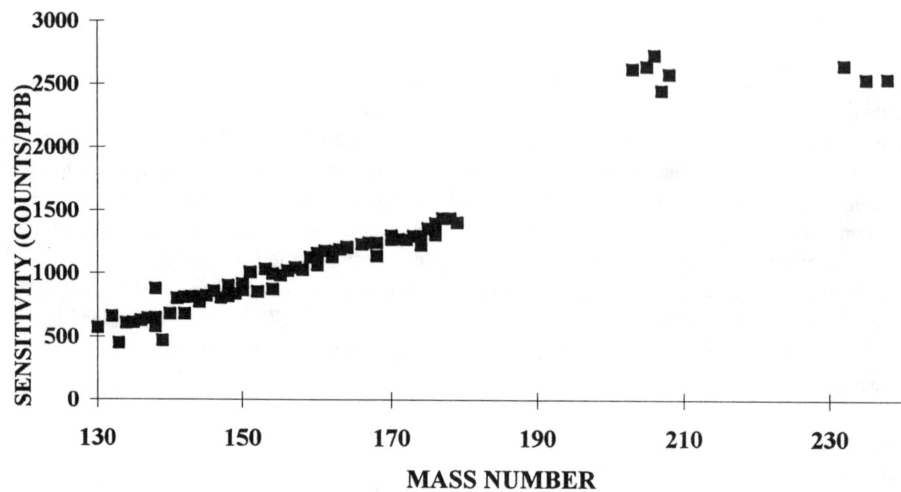

FIG. 1 - *ICPMS sensitivity versus mass for fission product and actinide regions.*

ICPMS was used in support of the Hanford Tank Waste Remediation Program. The goal of this program was testing and demonstration of waste processing strategies. Namely, pretreatment methods were developed for the removal of cesium, strontium, technetium, and transuranics from Hanford tank waste. Vitrification of the resultant low level waste was demonstrated as well as vitrification of the removed radionuclides combined with high level waste sludge. Radionuclide information from ICPMS was provided for each step of the development process. Strict program deadlines required rapid sample turnaround times which precluded lengthy sample pretreatment.

Figure 2 shows a portion of the ICPMS spectrum obtained for the untreated Hanford tank material diluted 2X in the ICPMS laboratory. This spectrum is used to demonstrate the analysis strategy used to meet the strict time requirements of this program.

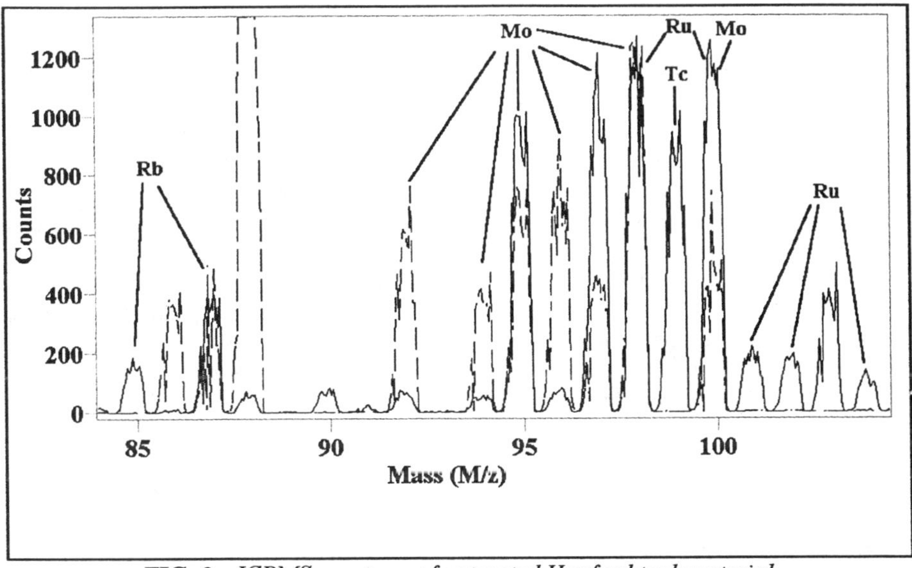

FIG. 2 - *ICPMS spectrum of untreated Hanford tank material.*
50 ppb standard---------, sample ———

In the mass region between mass 85 and mass 104, non-natural abundance distributions are measured for rubidium, molybdenum, and ruthenium. Likewise masses at 135 and 137 do not correspond to the natural abundance distribution for barium. Because of these anomalies in mass distributions, the standard software provided by the vendor cannot easily provide concentration data for the masses under investigation. Concentration data is obtained using the following strategy. The Quickbasic program, described previously in reference 2, extracts mass spectral data from the instrument generated binary files. The data is then imported into EXCEL for archival and quantitative analysis purposes and into GRAMS/32 for visual examination and comparison.

An Excel macro, developed in house and written in "Visual Basic for Applications", performs quantitative analysis on the ICPMS spectral data. The software detects the presence of fission products and actinides and flags the user using the the following key indicators:

Rubidium	mass 87>mass 85 and mass 88
Cesium	mass 137>mass 138
enriched Uranium	masses 234,235,236, and 237 present
Plutonium 238	mass 234>mass 235

The program calculates the concentration of the actinides, estimates the isobaric concentrations of fission species, and also calculates elemental concentrations based on natural abundances. For fission products, slope sensitivities are calculated based on isotopic abundances of possible standard mass interferences. Slopes for uncalibrated masses are estimated from a linear regression of known sensitivities on mass. The concentration at each mass value is then calculated. The program automatically prints the data in formatted sheets. An example of a portion of the data sheet is shown in Table 1. At the top of Table 1 is the diagnostics that identify potential fission products and actinides present in the sample. Elements and their corresponding natural abundance at a particular mass number are indicated as well as expected fission product and fission product abundance[*4,5*] for that mass number. Sensitivities for the mass number are listed and are highlighted on the computer display if they have been estimated from neighboring mass sensitivities. The standards and samples are identified and the measured isotope concentrations are listed. The raw counting data (not shown in table 1) is preserved in the spreadsheet so that analytical interferences and anomalies can be easily investigated. Plots of the internal standards are automatically generated so that any aberrations, such as signal suppression, can be readily observed.

Conclusion

Experimental evidence has been presented that supports the assumption of using a linear regression of neighboring mass sensitivities to calculate fission product and actinide concentrations for which certified standards are not readily available. An EXCEL macro is available that automates ICPMS data analysis of fission products and actinides using linear regression of calibrated neighboring mass sensitivities. The efficiency gained in the data analysis of ICPMS spectra allowed us to meet strict time requirements while providing quality data in support of the Hanford Tank Waste Remediation Program.

TABLE 1 - *Sample EXCEL macro printout of fission product calculations. Concentrations are in ppb*

Fission product diagnostics — 90849-10 fission Rb indicated, Tc indicated, fission Ru indicated, fission Cs indicated, enriched U indicated

	Element	Zr	Nb	Zr	Mo	Mo	Mo
	% abund.	17.1	100	17.5	15.9	16.7	9.5
	Element	Mo		Mo		Ru	
	% abund.	14.8		9.1		5.5	
	Element					Zr	
	% abund.					2.8	
	Fiss. Prod.	Zr*	Zr+	Zr*	Mo*	Zr*	Mo*
	Fiss. Yield	6.03	6.37	6.5	6.5	6.3	5.98
	Sensitivity	313	312	311	309	316	307
Sample	**Dilution Factor**	**92**	**93**	**94**	**95**	**96**	**97**
blank 1	1						
std1 1	1	*0.15*	*0.00*	*0.10*	*0.18*	*0.15*	*0.12*
std10 1	1	1.56	*0.02*	0.88	1.91	1.98	1.06
std100 1	1	14.89		9.82	17.89	18.70	12.16
std500 1	1	66.99	*0.03*	44.19	78.69	83.71	51.00
90849-10	10	93	*0.68*	76	1365	86	

	Element	Mo	Ru	Ru	Ru	Ru	Rh	Ru
	% abund.	24.4	12.7	12.6	17.1	31.6	100	18.6
	Element	Ru		Mo		Pd		Pd
	% abund.	1.9		9.6		1		11
	Element							
	% abund.							
	Fiss. Prod.	Mo*	Tc+	Mo*	Ru*	Ru*	Rh*	Ru*
	Fiss. Yield	5.78	6.1	6.28	5.18	4.29	3.03	1.88
	Sensitivity	338	305	304	303	302	300	299
Sample	**Dilution Factor**	**98**	**99**	**100**	**101**	**102**	**103**	**104**
blank 1	1							
std1 1	1	0.32	*0.00*	*0.15*	*0.07*	*0.01*		*0.10*
std10 1	1	2.88	*0.00*	1.33	*0.01*	*0.03*		*0.02*
std100 1	1	28.15	*0.02*	13.17		*0.01*		*0.04*
std500 1	1	122.37	*0.01*	57.87		*0.00*		*0.11*
90849-10	10	1391	1250	1674	289	255	562	162

References

[1] Kinard, W.F., Bibler, N.E., Coleman, C.J., and Wyrick, S.B., "Inductively Coupled Plasma-Mass Spectrometry Studies of the Chemistry of Fission Products and Actinides in High Level Wastes: Lessons That Can Be Applied to Environmental Measurements," *Radiochimica Acta*, Vols. 66/67, 1994, pp. 259-263.

[2] Kinard, W.F., Bibler, N.E., Coleman, C.J., Dewberry, R.A., Boyce, W.T., and Wyrick, S.B., "Applications of Inductively Coupled Plasma-Mass Spectrometry to the Determination of Actinides and Fission Products in High Level Radioactive Wastes at the Savannah River Site," *Applications of Inductively Coupled Plasma-Mass Spectrometry to Radionuclide Determinations, ASTM STP 1291*, R. W. Morrow and J. S. Crain, Eds., American Society for Testing and Materials, West Conshohocken, 1995, pp. 48-58.

[3] Kopajtic, Z., Rollin, S., Wernli, B., Hochstrasser, C., Ledergerber, G., and Jurcek, P., "Determination of Trace Element Impurities in Nuclear Materials by Inductively Coupled Plasma Mass Spectrometry in a Glove-box," *Journal of Analytical Atomic Spectrometry*, Vol. 10, 1995, pp. 947-953.

[4] Walker, F. W., Parrington, J. R., Feiner, F., *Nuclides and Isotopes*, 14th edition, General Electric Co., San Jose, CA, 1989.

[5] Bibler, N.E., Coleman, C. J., Kinard, W. F., "Relative Yields of U-235 Fission Products Measured in a High Level Rioactive Sludge at the Savannah River Site," *Proceedings*, "Spectrum 92," Nuclear and Hazardous Waste Management International Topical Meeting, , Boise, ID, 26 August 1992, p. 952

[6] Farmer, O.T. III, Reeves, J. H., Wyse, E. J., Clemetson, C. J., Barinaga, C. J., Smith, M. R., Koppenaal, D. W., "Analysis of Radioactive Waste Samples by Ion Chromatography-beta-ICP-MS," *Applications of Inductively Coupled Plasma-Mass Spectrometry to Radionuclide Determinations, ASTM STP 1291*, R. W. Morrow and J. S. Crain, Eds., American Society for Testing and Materials, West Conshohocken, 1995, pp. 38-47.

[7] Farmer, O. T. III, Smith, M. R., Wyse, E. J., Barinage, C. J., Koppenaal, D. W., "Separation and Quantitation of Radionuclides in Hanford Environmental and Waste Tank Samples Using IC-ICP/MS Techniques," *1996 Winter Conference on Plasma Spectrochemistry*, University of Massachusetts, Amherst, MA, 1996, p. 96.

Vernon D. Jones [1]

EVALUATION OF NEBULIZER PERFORMANCE WITHIN THE ICP-MS MEASUREMENT SYSTEM FOR ANALYSIS OF SRS RADIOLOGICAL WASTE TANK SIMULATED SOLUTIONS

REFERENCE: Jones, V.D., **"Evaluation of Nebulizer Performance within the ICP-MS Measurement System for Analysis of SRS Radiological Waste Tank Simulated Solutions,"** *Applications of Inductively Coupled Plasma-Mass Spectrometry to Radionuclide Determinations: Second Volume, ASTM STP 1344,* R.W. Morrow and J.S. Crain, Eds., American Society for Testing and Materials, 1998.

ABSTRACT: High level radioactive waste tanks at the Savannah River Site are high in salt content. The average Total Dissolved Solids (TDS) content is approximately 25%. For ICP-MS optimum signal stability and to reduce blockage of nebulizers and sampling orifices, it is usual to limit analyte solutions to a TDS content of nominally < 0.2%. Dilution to this level to reduce the matrix effect may push some analytes of interest below detectable levels. Five commercially available nebulizers were evaluated in a field study as part of the ICP-MS measurement system for their performance in a high salt matrix. The nebulizers surveyed were a meinhard concentric, cross-flow, micro-concentric (MCN), V-groove, and a direct injection nebulizer (DIN). Analytes spiked into non-radioactive diluted salt solutions ranging from nominal 0.25 - 1.0 % TDS were repetitively analyzed with the goal of determining stability of response signal and magnitude of any signal loss/suppression resulting from the diluted salt matrix. The cross-flow nebulizer provided the most stable signal for all salt matrices with the smallest signal loss/suppression due to this matrix. The DIN exhibited a serious lack of tolerance for TDS; possibly due to physical de-tuning of the nebulizer efficiency.

KEYWORDS: inductively coupled plasma mass spectrometry, high salt matrix, meinhard concentric nebulizer, cross flow nebulizer, micro-concentric nebulizer, V-groove nebulizer, direct injection nebulizer, signal loss and/or suppression

---

The Savannah River Site (SRS) located in Aiken, SC is operated by Westinghouse Savannah River Company under contract with the U.S. Department of Energy. Some of the waste currently being stored at the SRS in large underground tanks is high level

[1] Principal chemist, Westinghouse Savannah River Company, Technical Services Division, Building 772-F, Aiken, SC. 29808.

radioactive salt solutions. The average Total Dissolved Solids (TDS) content is approximately 25%. For ICP-MS (Inductively Coupled Plasma - Mass Spectrometry), it is usual to limit analyte solutions to a TDS content of nominally < 0.2% [1]. Dilution to this level may push some analytes of interest at or below detectable levels. Such as the analysis of low ppb Cu at m/z 65 after a 125 fold dilution with an interference at m/z 63.

The five nebulizers were evaluated for their effect on signal loss/suppression and precision of analyte signal in the presence of salt matrices. This evaluation was performed on a Hewlett Packard 4500 using non-radioactive simulated dilute high level waste tank solutions ranging in approximate dissolved solids content of 0.25 - 1.0 %. The purpose of this field study was to determine the best nebulizer for application within the ICP-MS system for this matrix based upon an acceptable signal reduction/suppression at the minimum dilution. An acceptable level of signal reduction/suppression for an internal standard was taken to be ≤ 40% based upon EPA method 200.8 [2].

Most nebulizer evaluations in the literature are carried out individually with comparison to another conventional pneumatic nebulizer (i.e., concentric, cross-flow or V-groove) . To the author's knowledge, there has been no published study of the 3 most widely used nebulizers and 2 specialty nebulizers on high salt matrices on the same instrument and operating conditions. This study seeks to underpin the knowledge base of several commercially available nebulizers for ICP-MS with specific focus on the "bottom line", how well do they work for practical application on the analysis of high salt matrices for radioactive waste characterization.

Experimental Method

Test Procedure

An HP-4500 ICP-MS (Hewlett Packard, Wilmington, DE, USA), was used to analyze the control standard and matrix test samples. The quadrupole mass analyzer was operated in the peak hopping mode while ion signals were detected by either pulse counting or analog signal dependent upon intensity; with an integration time of 0.33 sec. each and 3 repetitions per mass. A pulse to analog signal calibration (P/A factor) was performed prior to data acquisition using a 100 μg/L control standard at the masses designated for analysis.

The five nebulizers and accessories used were: a meinhard-type (MN) concentric (Meinhard, Santa Ana, CA, USA); cross-flow (CFN), V-groove (VGN) from Hewlett Packard; microconcentric (MCN) and direct-injection (DIN) with autosampler (CETAC Technologies, Omaha, NE, USA). An ASX-500 autosampler (also supplied by CETAC) was used with the following sequence: 60 sec. sample flush, 30 sec. equilibration, and following analysis, a 60 sec. rinse in 1% HNO_3. An ASX-100 autosampler was used with the MCN.

To reduce the number of variables, the plasma parameters were minimally adjusted for optimum use of the individual nebulizers (Table 1). RF power was set at 1230 W (exception for CFN). Sample depth was set at 7 mm which was found by

Hedrick of Hewlett Packard [3] to be the optimum position for the HP 4500 with high solids. An increased sampling distance allows longer residence time in the plasma and thus greater ionization while shorter sampling distances increase the sample and skimmer cone temperatures reducing the build up of material on cone surfaces. Carrier and blend gas flows as well as lens voltages were optimized for each nebulizer using a 10 ppb tuning solution of Li, Y, Ce and Tl after a minimum of 30 min. instrument warm-up.

TABLE 1 – *Instrument parameters.*

	MN	CFN	MCN	VGN	DIN
Date	Dec-96	Dec-96	Jan-97	Feb-97	Apr-97
Warm-up	>30 Min	>30 Min	>30 Min	>30 Min	>30 Min
RF Power	1230 W	1200 W	1230 W	1230 W	1230 W
Sample Depth	7 mm	7 mm	7 mm	7 mm	7 mm
Carrier Gas	1.2 L/Min	1.0 L/Min	0.94 L/Min	1.2 L/Min	0.35 L/Min
Blend Gas	0.0 L/Min	0.27 L/Min	0.39 L/Min	0.07 L/Min	0.0 L/Min
Peripump	0.1 rps	0.1 rps	0.1 rps	0.1 rps	0.0 rps
s/c Temp	2 oC	2 oC	2 oC	2 oC	NA
EM Voltage	(-)2010 V	(-)2010 V	(-)2010 V	(-)2010 V	(-)2100 V
Neb Press.					75 psi
GDP Press.					300 psi

The study solution was a simulated representation of the average waste tank matrix containing approximately 28% total dissolved solids but not including the actinide components found in the actual waste. Table 2 lists the contents of this salt solution. All the chemicals used to make the simulated waste were of analytical-reagent grade.

TABLE 2 – *Waste tank simulant.*

Average Waste Tank Simulant (~ 28% TDS)			
Na^+	5.00 M	SO_4^{2-}	0.13 M
OH^-	1.45 M	Cl^-	0.02 M
NO_3	1.86 M	F	0.01 M
NO_2^-	0.73 M	PO_4^{3-}	0.01 M
CO_3^{2-}	0.16 M	$C_2O_4^{2-}$	0.01 M
AlO_2^-	0.29 M		

For dilution of the simulated waste, aliquots were delivered with an electronic pipette (EDP Plus, Rainin Instruments, Woburn, MA, USA). Upon dilution, the test salt solutions contained 100 μg/L each of Sc, Y, In and Tb (as typical internal standards), and 1% v/v nitric acid (Fisher Scientific, OPTIMA grade, Pittsburgh, PA, USA) in 18 MΩ cm water (from this point forward called "dilute salt"). Three dilution schemes were used resulting in dilute salt TDS content of 0.28, 0.56, and 1.12%. An acid blank of 1% v/v HNO_3 also containing Sc, Y, In, and Tb at the 100 μg/L level was also prepared and will from this point forward be called the "control standard."

For each of the nebulizers, the control standard was analyzed several times to obtain a reference response for each analyte with no salt matrix present. The dilute salt solutions were then analyzed repetitively to measure the degree of matrix effect. Subsequent to the dilute salt analyses, the control standard was again analyzed several times to look at system recovery and instrument drift over the course of the analyses. Nebulizer data were tabulated for precision of dilute salt analyses and % drop in average signal due to the combined matrix effect of sample loss and signal suppression. This study did not attempt to differentiate between signal loss and suppression.

All of the data are represented graphically. For summary statistics, terbium was selected overall to compare the nebulizers for the possible application of actinide analyses in salt matrix. Of the standards selected, terbiums ionization energy is closest to U (565 vs. 584 kJ/mol [4] respectively).

Signal drop is expressed as

$$\%D = [(SM_{avg.} - CT_{avg.}) / CT_{avg.}]\, 100 \tag{1}$$

where

$\%D$ = percent signal drop
$SM_{avg.}$ = salt matrix average response
$CT_{avg.}$ = control standard average response

Downward instrument drift was evident in the control standard analyses and was not corrected for through the use of internal standards. Instrument response drift was conservatively compensated for by averaging the control standard analyses that bracket the salt matrix analyses. This "time averaged" control standard response was then used to calculate the average percent signal drop due to the matrix. It is understood that this number could be biased low if partial blockage of the nebulizer or sample cone has occurred during the course of the analyses. However, the slope of the control standard analyses before the matrix analyses gives an extrapolated indication of where the control acid analyses after running the salt matrix should be due to instrument drift alone.

Actual radioactive waste tank analyses may result in greater signal suppression of low m/z ions than summarized due to the presence of actinide high m/z species giving rise to "space charge effects".

Results

Meinhard Concentric

The meinhard concentric nebulizer used was the "type A" which is not the optimum for high solids. Figure 1 shows the signal responses for analysis of 0.28%, 0.56%, and 1.12% TDS salt matrices; each bracketed by analyses of the control standard. All analyses were performed on the same day. Twenty-five replicate analyses were performed for each salt matrix. The 0.28% TDS matrix was the only one evaluated that had $\leq$ 40% signal drop using the type A meinhard nebulizer.

FIG. 1 – *Meinhard concentric high TDS response.*

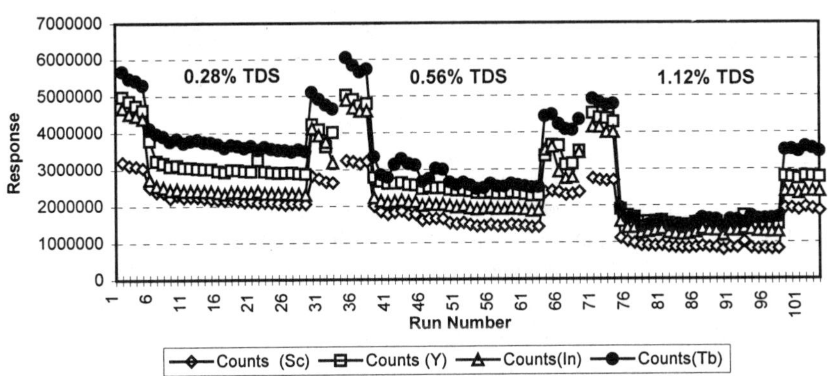

TABLE 3 – *Meinhard concentric data summary for Tb.*

Data Summary	Tb, 0.28% TDS	Tb, 0.56% TDS	Tb, 1.12% TDS
Average	3698290	2756485	1582921
STD @95% CL	331953	575057	255561
RSD @95% CL	9.00%	20.90%	16.10%
% Signal	-28.40%	-43.66%	-60.82%

Cross-Flow Nebulizer

For the cross-flow nebulizer, the RF power was reduced to 1200 W as this produced a more stable response signal. Figure 2 shows the signal responses for analysis of 0.28%, 0.56%, and 1.12% TDS salt matrices; each bracketed by analyses of the control standard. Analyses were performed on 3 non-consecutive days. Twenty replicate analyses were performed for each salt matrix. All of the dilute salt matrices had acceptable signal intensity (≤ 40% drop) using the cross-flow nebulizer.

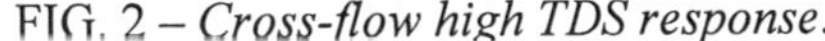
FIG. 2 – *Cross-flow high TDS response.*

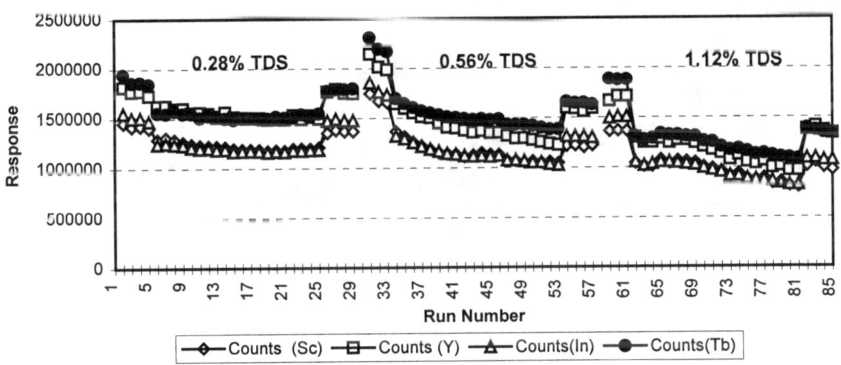

TABLE 4 – *Cross-flow data summary for Tb.*

Data Summary	Tb, 0.28% TDS	Tb, 0.56% TDS	Tb, 1.12% TDS
Average	1523650	1496159	1218408
STD @95% CL	52064	178580	186900
RSD @95% CL	3.42%	11.94%	15.34%
% Signal	-16.8%	-21.0%	-23.1%

Micro-Concentric Nebulizer

The micro-concentric nebulizer model MCN-100 made by CETAC, is similar to the meinhard concentric nebulizer except that the diameter of the capillary and gap for the Ar carrier gas are smaller. The unique characteristics of the MCN-100 are a low sample uptake rate of about 20 μL/min and high efficiency. Higher Ar gas pressure and the design of the nebulizer tip may help prevent salt deposition. The peristaltic pump tubing supplied by CETAC with the MCN is made of PVC with a 0.19 mm ID. For ease of assembly and use, PVC tubing with a 0.25 mm ID was substituted. The PVC pump-tubing tends to elongate during the increased rpm for the rinse cycle and then relax to normal length during pump stabilization and analysis after pump speed is reduced. To facilitate using the increased pump speed for shorter rinse out times, the pump tubing was stretched to the longest stops on the pump head to minimize variations in length. The stretching and subsequent shrinking of the pump tubing also varies the internal diameter and resulting sample flow rate. This may contribute to some of the variability of the MCN technique. Using a longer rinse time (X5) and constant pump speed could possibly improve the precision of the results summarized below. Figure 3 shows the signal responses for analysis of 0.28%, 0.56%, and 1.12% TDS salt matrices; each bracketed by analyses of the control standard. The first three analysis sets were performed on the same day. The last run of 1.12% TDS salt matrix was analyzed on the next day to confirm the signal drop off of both salt and final control standard. Twenty replicate analyses were performed for each salt matrix. Only the 0.28% TDS matrix exhibited ≤ 40% signal drop using the MCN.

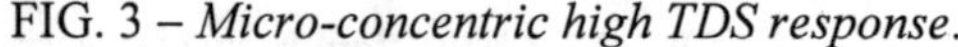
FIG. 3 – *Micro-concentric high TDS response.*

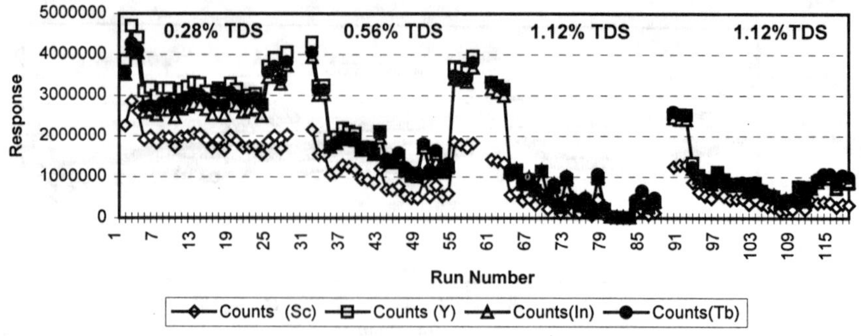

Table 5 – *Micro-concentric nebulizer analyses for Tb.*

Data Summary	Tb, 0.28% TDS	Tb, 0.56% TDS	Tb, 1.12% TDS	Tb, 1.12% TDS
Average	2873803	1580803	568979	762730
STD @95% CL	279949	655497	858546	492909
RSD @95% CL	9.74%	41.47%	150.9%	64.62%
% Signal	-23.9%	-54.8%	-82.4%*	-70.0%*

**[Note: % signal drop for Tb in 1.12% TDS based on 3 initial control standard analyses prior to salt matrix.]*

V-Groove Nebulizer

The VGN used was that supplied with the HP 4500 instrument. Figure 4 shows the signal responses for analysis of 0.28%, 0.56%, and 1.12% TDS salt matrices, each bracketed by analyses of control standard. In addition to Sc, Y, In, and Tb; the elements Th and U were also added at the 100 µg/L level. All analyses were performed on the same day. Twenty replicate analyses were performed for each salt matrix. The 0.28% and 0.56% TDS matrices exhibited ≤ 40% signal drop using the V-groove nebulizer.

FIG. 4 – *V-groove high TDS response.*

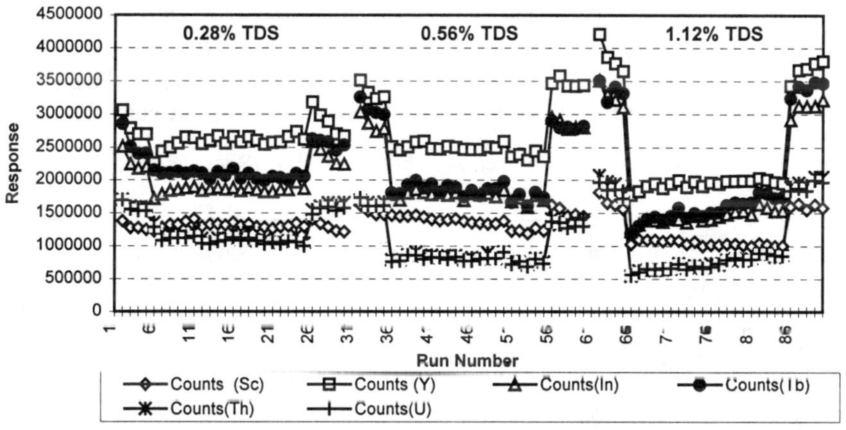

TABLE 6 – *V-groove data summary for Tb and U.*

Data Summary	Tb, 0.28%	Tb, 0.56%	Tb, 1.12%	U, 0.28%	U, 0.56%	U, 1.12%
Average	2078037	1833196	1543179	1088016	795968	732833
STD @95% CL	107532	208048	345965	128789	106354	204329
RSD @95% CL	5.17%	11.35%	22.42%	11.84%	13.4%	27.88%
% Signal	-18.6%	-37.6%	-54.3%	-30.4%	-45.6%	-61.0%

Direct Injection Nebulizer

The direct injection nebulizer (DIN) model Microneb 2000 is manufactured by CETAC. When using the DIN, the sample is delivered through a narrow inert capillary tube. Argon gas passes into the nebulizer tip through the narrow annular space between the sample capillary and the nebulizer tip. The result is a fine aerosol mist with a narrow particle size distribution. Due to the absence of large droplets, a spray chamber is not needed and the sample is directly injected into the central channel of the ICP. Because sample liquids must pass through a long narrow capillary, a sample injection valve is coupled with a high pressure gas displacement pump to deliver the sample to the nebulizer. In addition to control of gas flows and pressures, the DIN aerosol is optimized through extending or retracting the capillary tube relative to the nebulizer tip. This adjustment is under stepper motor control. One of the advertised advantages of the DIN include small sample sizes and quick rinse-outs between samples. This translates into improved efficiency of operations and reduced consumables [5] as well as waste reduction; both of which translate into potential cost savings. Application of the DIN for use in United States Department of Energy (DOE) and contractor laboratories to minimize analytical wastes from radioactive and/or hazardous samples has previously been proposed [6].

Figure 5 shows the signal responses for analyses of 0.28%, 0.56%, and 1.12% TDS salt matrices; each bracketed by analyses of control standard. All analyses were completed consecutively in one day. After the first 2 analyses of 0.28% TDS salt, the signal begins an abrupt drop. After 12 consecutive analyses of the 0.28% TDS salt, the response signal does not recover when switched back to analyzing the control standard.

FIG. 5 – *Direct injection nebulizer high TDS response.*

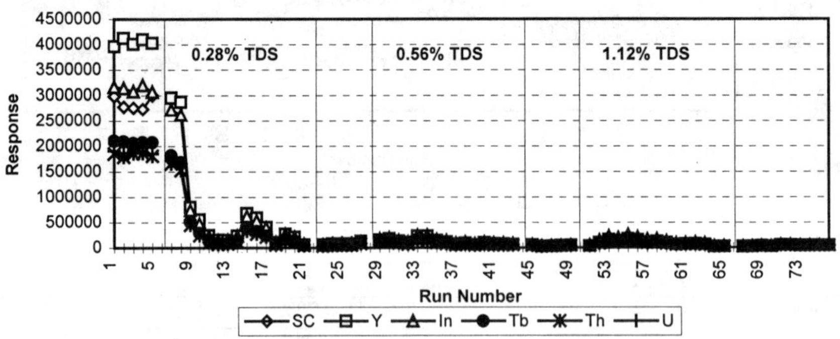

In attempting to regain instrument sensitivity, the sample and skimmer cones were replaced with new and the spectrometer and DIN re-tuned after running in the instrument for several hours. An apparent displacement of the capillary within the DIN proved to be the primary cause of the signal loss and was very sensitive and needed to be re-optimized each time the ICP was started. Although not represented here in a figure, the control and dilute salt matrix were again analyzed with the DIN. The analysis sequence was to analyze the control standard (5x) followed by repetitive analysis of a dilute salt matrix at 0.10% TDS. After the first 2 analyses of 0.10% TDS, the response dropped nearly to zero and remained low for the remainder of the 0.10% TDS. Further work with the DIN after these analyses gave further indication that it is extremely sensitive to the matrix TDS content. After a severe signal loss, the DIN could be re-tuned to regain sensitivity by adjusting the position of the capillary relative to the nebulizer tip and optimizing the vertical and horizontal torch positions. This was required frequently when running the dilute salt matrices

Conclusion

For analysis of simulated radioactive high salt matrices > 0.2% TDS, the cross-flow is the nebulizer of choice in our laboratory at the parameters studied for both stability of response and to minimize matrix effect on signal loss/suppression. Figure 6 shows the compiled results of all the nebulizers studied normalized to an initial control standard response of 5 million counts for Tb. As in the previous figures, the dilute salts 0.28%, 0.56% and 1.12% are bracketed by control standard analyses. The MCN exhibited severe signal loss/suppression at 1.12% TDS, while the DIN is not recommended for high solids work.

FIG. 6 – *Normalized high TDS response for all studied nebulizers.*

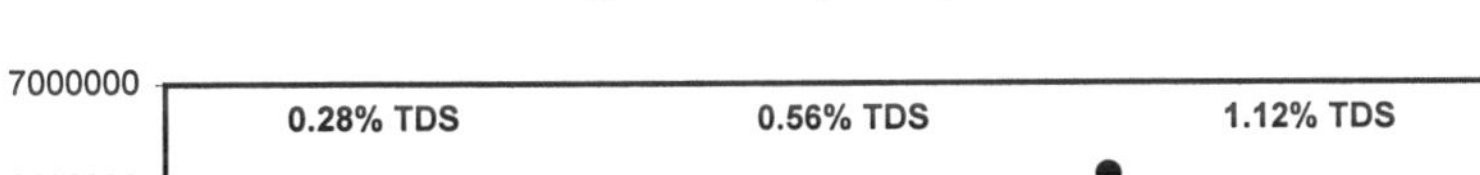
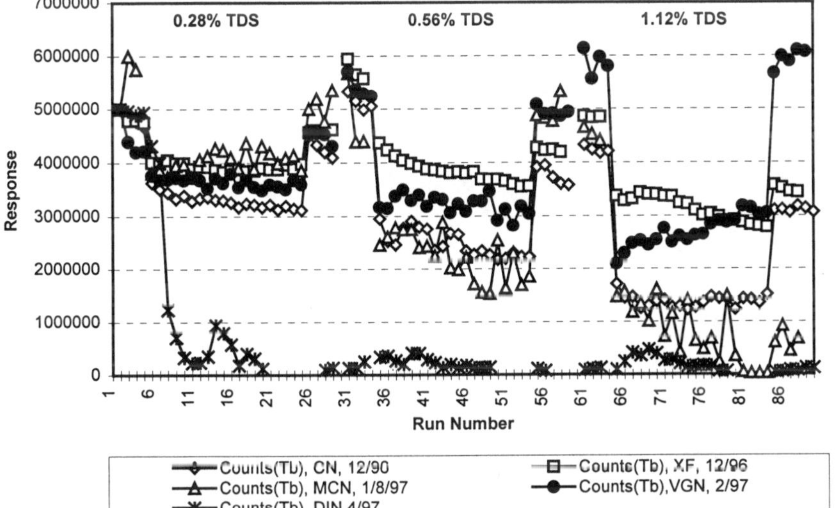

References

[1] Jarvis, Gray, Houk, *Hand Book of ICP-MS*, Section 5.3, Blackie, Glasgow, 1992.

[2] EPA Method 200.8; "Determination of trace elements in waters and wastes by inductively coupled plasma - mass spectrometry," EMMC version, revision 5.4. EMSL, Cincinnati, OH, May 1994.

[3] Hewlett Packard Application Note 228-320; "Analysis of Samples Containing High TDS Levels Using the HP 4500 ICP-MS," March 1995.

[4] Jarvis, Gray, Houk, *Hand Book of ICP-MS*, Appendix 2, Blackie, Glasgow, 1992.

[5] CETAC Application Note: 001DIN1196U.

[6] Crain, JS and Kiely, JT, J. Anal. At. Spectrom., Vol. 11, July 1996, pp. 525-527.

Philippe G. Bienvenu,[1] Eric A. Brochard,[1] and Emmanuel A. Excoffier[2]

DETERMINATION OF LONG-LIVED BETA EMITTERS IN NUCLEAR WASTE BY INDUCTIVELY COUPLED PLASMA-MASS SPECTROMETRY

REFERENCE: Bienvenu, P.G., Brochard, E.A., and Excoffier, E.A., **"Determination of Long-lived Beta Emitters in Nuclear Waste by Inductively Coupled Plasma - Mass Spectrometry,"** *Applications of Inductively Coupled Plasma - Mass Spectrometry to Radionuclide Determinations: Second Volume, ASTM STP 1344*, R.W. Morrow and J.S. Crain, Eds., American Society for Testing and Materials, 1998.

ABSTRACT: Inductively Coupled Plasma–Mass Spectrometry (ICP-MS) constitutes a very attractive alternative to radiochemical techniques for the determination of long-lived radionuclides considered critical for the safety of nuclear waste repositories. By measuring isotopic abundance instead of ionizing radiation, it is shown that ICP-MS should exhibit lower detection limits than liquid scintillation counting for the measurement of β emitters with half-lives longer than ~10^4 years. However, a specific preparative chemistry is generally required prior to measurement in order to separate the analyte from major radioactive contaminants and potentially interfering isotopes. Original methods we have recently developed for the determination of three long-lived β emitters: ^{93}Zr, ^{107}Pd and ^{135}Cs in radioactive waste are described in this paper. The procedures involve various separative and measurement strategies, such as liquid-liquid extraction, chromatographic separation as well as electrothermal vaporization coupled with ICP-MS. The performances are discussed in terms of selectivity and detection capabilities.

KEYWORDS : ICP-MS, radiochemical methods, liquid chromatography, sample preparation, zirconium-93, palladium-107, cesium-135

---

[1] Commissariat à l'Energie Atomique, DCC/DESD, CE Cadarache, F13108 Saint Paul Lez Durance, France.
[2] SGS Qualitest, DSN, 191, Av. A. Briand, F94237 Cachan Cedex, France.

The management of nuclear waste storage and disposal sites requires a reliable knowledge of the activity of all radionuclides present in waste packages, and more especially long-lived isotopes which will dominate environmental consequences in the future. Safety studies conducted in this field have led regulatory authorities to establish lists of such "critical" nuclides, with limitations regarding the maximum allowable activities per package as well as total radioactive capacity of repository sites. As an example, Table 1 gives the relevant specifications for β emitters in radwaste packages destined to the surface site in France.

Unfortunately, the determination of most critical nuclides is impossible by routine non-destructive techniques because of the nature of the principal radiation (mostly α, pure β or X-ray emission), the low specific activity and the low content of these isotopes in real waste samples. Destructive procedures involving discrete sampling and physico-chemical treatment have to be carried out prior to measurement. Regarding measurement techniques, the analyses of radioisotopes are generally performed using radiometric techniques, i.e. α-, X- and γ- spectrometry, or liquid scintillation counting. In the case of long-lived radionuclides, abundance mass spectrometry such as Inductively Coupled Plasma–Mass Spectrometry (ICP-MS) is expected to constitute a complementary tool to radiochemical methods, achieving lower detection limits in less time. Many recent publications have stressed this great potential [*1-4*]. Some applications of ICP-MS for the determination of three long-lived nuclides, ^{93}Zr, ^{107}Pd and ^{135}Cs, in radioactive waste samples are described in this paper, including chemical and/or physical pretreatment as well as measurement methodology.

TABLE 1—*Long-lived beta emitters considered to be critical for the safety of nuclear waste surface disposal in France.*

Radionuclide	Half-life, year	Origin	Limit of Acceptability, GBq/t
^{14}C	5.7x10^{3}	activation	200
^{59}Ni	7.5x10^{4}	activation	63
^{63}Ni	1.0x10^{2}	activation	12000
^{93}Zr	1.5x10^{6}	activation, fission	2.6
^{94}Nb	2.1x10^{4}	activation	0.12
^{99}Tc	2.1x10^{5}	fission	1
^{107}Pd	6.5x10^{6}	fission	370
^{129}I	1.6x10^{7}	fission	0.046
^{135}Cs	2.3x10^{6}	fission	10
^{151}Sm	9.3x10^{1}	fission	1600

ICP-MS: An Alternative to Radiometric Techniques for the Measurement of Long-Lived Radionuclides

Quantitative analysis of a radionuclide can be performed either by radiometric techniques that measure ionizing radiation generated by radioactive transmutations (activity, in Bq) or by mass spectrometric techniques, which measure abundance of the relevant isotope (mass, in g). From the basic law of radioactivity, it is easy to demonstrate that the activity A_t of a nuclide is proportional to the number N_t of radioactive isotopes at the time *t*:

$$\mathbf{A_t = \lambda . N_t} \qquad \lambda\text{: decay constant}$$

The number of atoms N_t can be expressed by means of the amount m_t of the isotope:

$$\mathbf{A_t = \lambda . m_t . \frac{N_A}{M}} \qquad N_A\text{: Avogadro constant; } M\text{: atomic mass}$$

Substituting half-life *T* for the decay constant λ, it follows that:

$$\mathbf{A_t = m_t . \frac{\ln 2}{T} . \frac{N_A}{M}}$$

This formula indicates that the activity of a radionuclide in a sample is all the more important when the half-life of this isotope is short and the atomic mass is small. Conversely, it means that the longer the half-life the lower the specific activity (A_t/m_t), i.e., the lower the activity per unit mass amount. Regarding measurement techniques, this observation implies that mass spectrometric techniques are thus expected to display higher performance figures than radiometric techniques in the determination of long-lived radionuclides.

On the basis of the detection limits achievable by each technique, it is possible to compare the relative capabilities of Inductively Coupled Plasma–Mass Spectrometry with Liquid Scintillation Counting for the measurement of β emitters. The diagram in Figure 1 enables us to define a "half life cutoff" between ICP-MS and LSC where the detection limit by ICP-MS is equal to the detection limit by LSC, which is between 3.10^3 years for the heaviest isotopes and 7.10^4 years for the lightest ones. Disregarding difficulties from sample matrices, this conclusion emphasizes the great potential of ICP-MS for the determination of long-lived isotopes.

Experimental

Instrumentation

The ICP-MS used throughout this work was an ELAN 6000 (Perkin Elmer/Sciex). Two types of sample introduction techniques were used: pneumatic nebulization with a cross-flow system and a Scott-type spray chamber, and electrothermal vaporization with a Massmann-type graphite furnace (HGA-600MS, Perkin Elmer). Typical instrument operating conditions and data acquisition parameters are given in Tables 2 and 3.

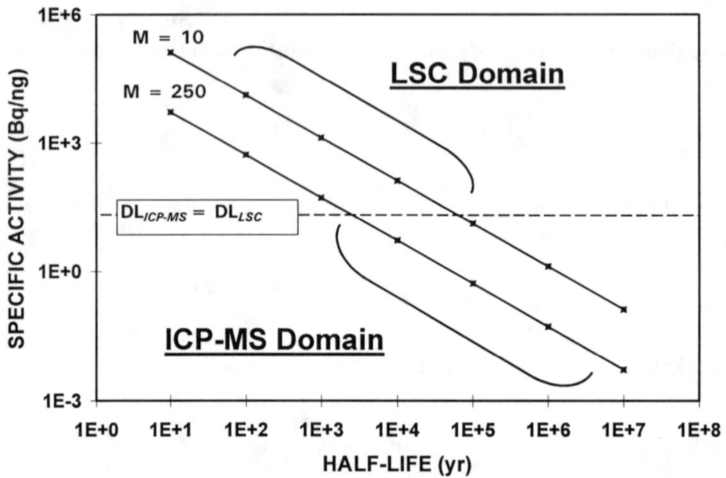

FIG. 1—*Specific activity (A_t/m_t) versus half-life for radionuclides with atomic masses ranging between 10 and 250. The half-life cutoff is calculated for instrumental detection limits (DL) equal to 1 ng/L for ICP-MS and 20 Bq/L for LSC.*

Reagents

All the reagents were of high analytical grade (Suprapur HNO_3, HCl, HF, NH_4OH, Merck). De-ionized water (18.2 MΩ.cm) was produced by a pure water generating system (Milli-Q Plus, Millipore).

Chromatographic resins used for Zr and Cs analyses, namely SM7 and AG1-X8 were from Bio-Rad. Tributylphosphate, triphenyl-phosphine and ethanol were from Merck. Pentyl-2 nitrophenyl ether (NPPE) is synthesized by Eras Labo (Lyon, France).

Calibration

The absence of certified solutions for ^{93}Zr, ^{107}Pd and ^{135}Cs leads us to calibrate the instrument using alternative stable isotopes from the same elements, ^{92}Zr, ^{105}Pd and ^{133}Cs respectively. Additionally, ^{103}Rh was added to samples and standards at 25 µg/L and used as internal standards. Standards solutions were prepared from dilution of 1g/L single element certified solutions (Merck).

Application of ICP-MS for the Characterization of Radioactive Waste Samples

Determination of Zr-93

Zirconium 93 is generated both by activation from fuel cladding and by fission reactions in nuclear reactors. Considering its long half-life (1.5 10^6 year), it is an interesting candidate for ICP-MS measurement. The principal isobaric interferences to be taken into account are ^{93}Nb, which is the only natural isotope of niobium, and to a lesser extent ^{93}Mo, which is a fission product with a half-life of 3.5 10^3 years.

TABLE 2—*ICP-MS operating conditions.*

RF power	1050 W
Plasma Ar flow	15 L/min
Auxiliary Ar flow	1.0 L/min
Nebulizer flow rate	0.9 L/min
Resolution	Normal

TABLE 3—*Data acquisition parameters.*

	Pneumatic Nebulization	Electrothermal Vaporization
Dwell time	50 ms	35 ms
Points across peak	1	1
Sweeps per reading	25	2
Readings per replicate	1	30
Number of replicates	10	3

The procedure developed for the determination of ^{93}Zr in radioactive waste samples consists first in an acidic dissolution with a mixture of HNO_3 – HCl – HF. A 0.5 g sub-sample is placed into a PTFE digestion vessel and 10 mL of concentrated acids is added. The mixture is then heated at 80°C in a drying oven for 12 hours. The solution is evaporated nearly to dryness and re-dissolved in 7N nitric acid. The selective extraction of Zr is performed by reversed phase chromatography with tributyl-phosphate as the static extractant, fixed on a polyacrylate support (BioRad SM7 resin). The affinity of Zr for TBP in nitric media has been described for a long time [*5*]. In the present work, Zr is complexed by TBP in 7N nitric solution, while most of the other elements are not retained and thus flowed down through the column. Subsequently, Zr is eluted with 2N nitric acid. The elution profiles in Figure 2 illustrate the great selectivity of the separation. Based on the concentrations of stable ^{92}Zr measured in the dissolved sample and in the 7N eluent, it follows that the recovery yield for Zr is around 90% (Table 4). Moreover, the separation factors with respect to both interfering isotopes and γ emitters are also very good since around 100% of these nuclides are eliminated in the 2N eluent (Table 4). Measurement of ^{93}Zr is carried out using conventional ICP-MS parameters with nebulization sampling. The calculation of the instrumental detection limit, considering 3σ blank, gives a value of 0.4 ng/L, that is 0.04 Bq/L. For an initial sample weight of 0.5 g, this value corresponds to a limit of 10^{-3} Bq/g (~10^{-11} g/g).

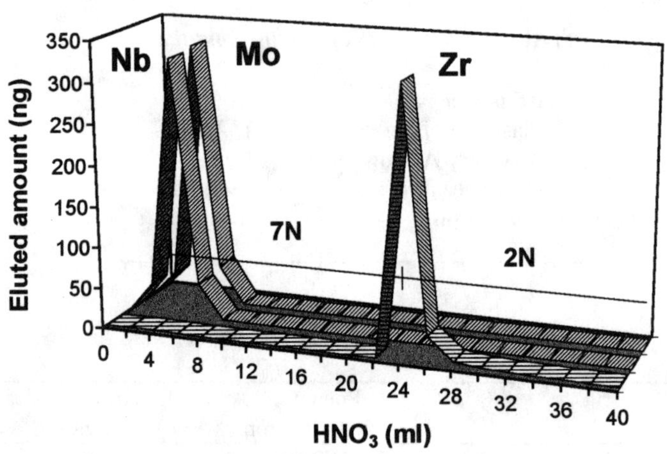

FIG. 2—*Elution profiles of a synthetic solution with 400 pg Nb, Zr and Mo.*

TABLE 4—*Zr, Nb, Mo contents and ^{60}Co activity in dissolved sample, HNO_3 7N and HNO_3 2N chromatographic fractions from a water filter sample.*

	Dissolution	7N Elution	2N Elution	Yield, %
Zr µg	400	20	360	90
Nb µg	0.80	0.80	...	<1
Mo µg	2.1	2.1	...	<1
^{60}Co KBq	320	330	...	<1

Determination of Pd-107

Palladium 107 is a fission product with a half-life of 6.5 10^6 years. The measurement of this isotope by ICP-MS suffers from an interference with ^{107}Ag, which represents 52% of natural Ag.

The dissolution procedure used for the determination of ^{107}Pd in solid samples is similar to the procedure described previously for ^{93}Zr. The subsequent separation step was first developed by Tardy [*6*] in our laboratory for the measurement of ^{107}Pd in nitric effluents from reprocessing plants. It relies on liquid-liquid extraction from 2N HNO_3 solution with triphenyl-phosphine diluted in nitro-2-phenyl pentyl ether (NPPE). Pd is then selectively back-extracted in 4M ammonia. This procedure has been tested on solid waste samples such as spent resins and water filters from nuclear reactors. The results in Table 5 show that the chemical treatment enables us to achieve an excellent decontamination factor with respect to γ emitters, as exemplified by ^{60}Co data, and a rather good separation factor between Pd and Ag. Unfortunately, the chemical yield calculated from stable ^{105}Pd ranges only from 30 to 40% at the best (Table 5).

The application of this method turned out to be efficient for the quantification of ^{107}Pd in most waste samples, using routine conditions for ICP-MS measurement. Nevertheless, it has been inadequate for the characterization of peculiar samples, with very high Ag contents, probably originating from control rod failures. Although more than 99% of Ag was removed from the initial sample by liquid-liquid extraction, the residual amount in the back-extracted phase was high enough to induce a substantial interference of ^{107}Ag on ^{107}Pd signal. The intensities reported in Table 6 indicate that the huge signal at mass 107 is predominantly due to the ^{107}Ag isotope, as demonstrated by the coherent signal at mass 109 due to the ^{109}Ag isotope.

TABLE 5—*Pd, Ag contents and ^{60}Co activity in dissolution and back-extraction phases from a water filter sample. The amount of Pd was introduced into the dissolved sample as internal standard.*

	Dissolved Sample	Back extracted Phase	Recovery Yield, %
Pd μg	5.0	1.8	36
Ag μg	87	0.2	0.2
^{60}Co kBq	1970	<10	<1

In order to perform a more complete elimination of Ag, we investigated the feasibility of electrothermal vaporization (ETV) coupled with ICP-MS analysis. The definition of an appropriate thermal program allowed us to separate Pd from Ag through a fractional volatilization process. The curves in Figure 3 show that Ag is readily eliminated at temperatures below 1000°C while Pd starts to volatilize above 1500°C. Table 7 gives the sampling program developed for this determination. After introduction of 50 µL solution into the furnace, the program consists first of a drying step at 120°C to remove water, followed by a volatilization step at 1500°C to remove Ag. During this time, the gas flow promotes the complete purging of volatile species from the furnace. The ETV cell is then sealed and the gas stream flows through the furnace towards the plasma. The temperature is ramped up very quickly and stabilized at 2700°C during 7 seconds to volatilize Pd. Systematically, the system is cleaned between two analyses by heating at 3000°C.

The selectivity of ^{107}Pd measurement on real samples was controlled by measurement of both masses 107 and 109 intensities. Considering the isotopic composition of Ag, the results in Table 6 allow us to calculate that the interference from ^{107}Ag would represent less than 5% of the signal at mass 107.

Although the detection limit (~10 ng/L) is a little higher when compared with nebulization sampling, ETV-ICP-MS was quite successful in terms of selectivity for measuring ^{107}Pd. The procedure makes it possible to determine ^{107}Pd in real waste sample at activity levels down to 10^{-2} Bq/g (~10^{-10} g/g). The analytical precision calculated for 3 replicates per sample was less than 10% relative for activity levels around 20 Bq/L.

TABLE 6—*Intensities measured at masses 107 and 109 on the back-extracted phase of a water filter sample, using pneumatic nebulization and ETV sampling.*

	Intensity, cps	
	Mass 107	Mass 109
Pneumatic Nebulization	11959	10566
Electrothermal Vaporization	4644	179

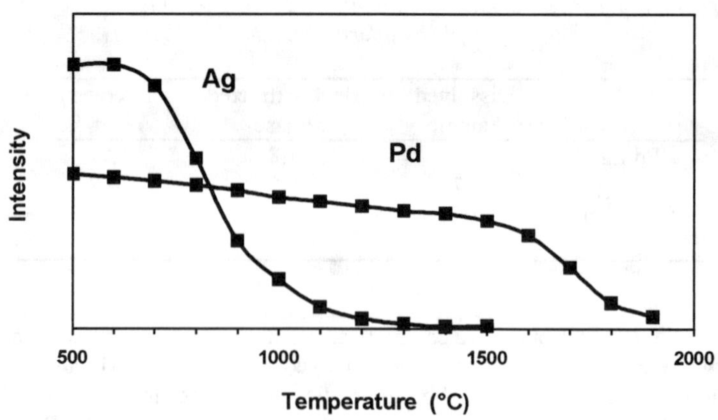

FIG. 3—*Thermal profiles on synthetic solutions with 1 µg/L Pd and Ag.*

TABLE 7—*Thermal program for ETV-ICP-MS analysis of ^{107}Pd.*

Step	Temperature, °C	Ramp time, sec	Hold time, sec
1- Drying	120	12	75
2- Ag Volatilization	1500	15	120
3- Pd Volatilization	2700	0	7
4- Cleaning	3000	5	5

Determination of Cs-135

Cesium-135 is one of the most difficult nuclides to determine using radiometric techniques. In fact, the analysis of ^{135}Cs by liquid scintillation counting is interfered by the emission of ^{137}Cs, another pure β emitter with a much higher specific activity ($t_{1/2}$ = 30 year), which is impossible to separate prior to the measurement. Therefore, mass-spectrometric techniques are the only means to determine ^{135}Cs in real waste samples. The main difficulty for isotopic measurements comes from the isobaric interference of ^{135}Ba, which represents 6.6% of natural barium. In this work, the separation of Cs from Ba has been carried out by anion exchange chromatography, following the study by Shabana and Ruf [*7*]. After partial dissolution of samples with 2N nitric acid, 75% ethanol is added to the solution to increase the partition coefficient of Ba relative to Cs. In these conditions, Ba is strongly retained on the column (BioRad AG1X8 resin) while Cs is rapidly eluted by the above mixture (Figure 4). The recovery yield deduced from ^{137}Cs activities is around 70% (Table 8). The decontamination factor with respect to major contaminants is, however, very low since most of the γ emitters are eluted together with Cs. Consequently, the radioactivity of the solution can render the analysis impossible using routine ICP-MS conditions due to irradiation effects as well as contamination problems.

Once again, the application of electrothermal vaporization in conjunction with ICP-MS constituted a successful alternative to avoid these difficulties by using very small volumes of solution (50 μL). In addition to this, the definition of an appropriate thermal program (Table 9) permits us to assure a further separation of Cs from residual traces of Ba. Figure 5 shows that the volatilization of Ba occurs at a much higher temperature than Cs. This way, it is possible to volatilize and to measure selectively ^{135}Cs, while the majority of Ba remains in the furnace.

In a previous work [*8*], ^{135}Cs was determined by ETV-ICP-MS using external calibration with ^{133}Cs standard solutions. Optimized operating conditions permitted us to achieve detection limits close to 10^{-3} Bq/L (10 pg/L). In practice, analyses at such low levels proved to be very delicate because of pollution problems concerning the natural Cs isotope used for the calibration. To overcome these difficulties, a much more reliable method has been applied in the present study. The quantification of ^{135}Cs was made by simultaneously measuring ^{135}Cs and ^{137}Cs intensities using ETV-ICP-MS. The concomitant presence of ^{137}Cs at substantial levels in the samples is really interesting since the two isotopes exactly behave in the same manner all along the analytical procedure, making ^{137}Cs the tracer of choice for this determination. As far as the ^{137}Cs activity is systematically determined in the initial bulk sample by non-destructive γ spectrometry (from ^{137m}Ba activity), the final measurement of ^{135}Cs/^{137}Cs ratio in the extracted solution allows us to calculate the initial ^{135}Cs activity in the waste sample.

Results reported in Table 10 for four spent resin samples show the reproducibility of the procedure. Despite the large variability of Cs abundance, ^{135}Cs/^{137}Cs ratios are remarkably coherent. The accuracy of the analytical procedure is supported by the fact that the measured ratio in waste is identical to the value calculated from spent fuel models, that is 5.10^{-6}

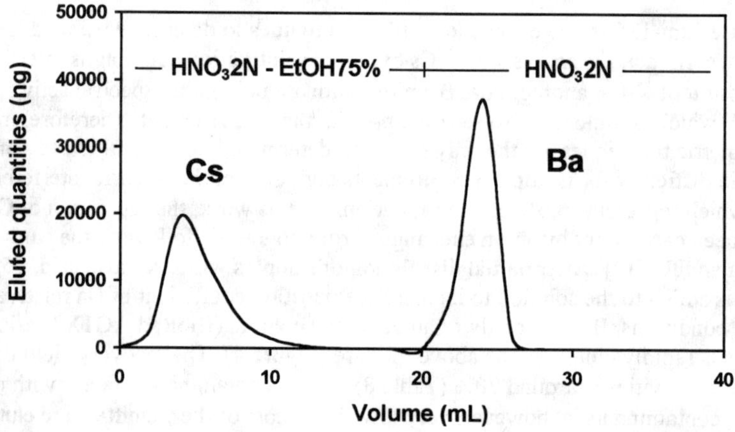

FIG. 4— *Elution profiles of a synthetic solution with 0.5 μg Cs and Ba.*

TABLE 8—*Balance of ^{137}Cs for dissolution and chromatographic separation steps for the analyses of resin samples.*

Sample Number	Resin 1	Resin 2	Resin 3	Resin 4
Sample weight, g	1.08	0.64	0.85	0.90
Initial solid sample, KBq	0.88	2.5	24	36
Dissolved sample, KBq	0.66	2.2	18	32
Eluted fraction, KBq	0.63	1.8	15	26
Recovery yield, %	72	73	62	74

TABLE 9—Thermal program for ETV-ICP-MS analysis of ^{135}Cs.

Step	Temperature, °C	Ramp time, sec	Hold time, sec
1- Drying	120	12	100
2- Ashing	1000	5	10
3- Cs volatilization	1800	0	5
4- Cleaning	3000	5	5

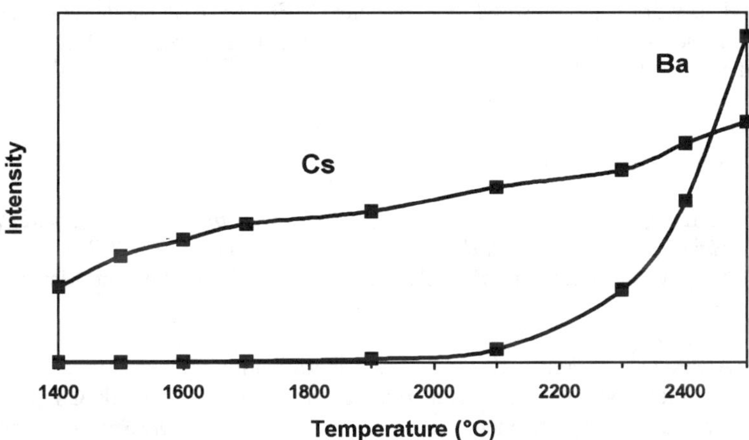

FIG. 5—*Thermal profiles on synthetic solutions with 1 μg/L Cs and Ba.*

TABLE 10—^{135}Cs and ^{137}Cs intensities measured by ETV-ICP-MS in resin samples.

Sample Number	^{135}Cs, cps	^{137}Cs, cps	$^{135}Cs/^{137}Cs$
Resin 1	10197	26563	0,38
Resin 2	89998	256500	0,35
Resin 3	147683	416670	0,35
Resin 4	3782	10637	0,36

Conclusion

ICP-MS turns out to be a very powerful tool for the determination of critical long-lived radioisotopes in nuclear waste. However, it is evident that intrinsic detection capabilities of ICP-MS can only be reached when applying very specific separation procedures prior to measurement. Besides chemical schemes involving organic extractants, the great potential of thermal separations with ETV-ICP-MS has been stressed in this work. Even though new ICP-MS instruments are expected to display higher and higher performances in terms of both sensitivity and selectivity, the development of pretreatment strategies will remain indispensable to eliminate isobaric interferences until extreme resolution factor has been reached.

Acknowledgment

This work was sponsored by Electricité De France (EDF) under contract CT6/FT2604-FA6720. The authors would like to thank A. Hérès and D. Révy for helpful comments.

References

[1] Morrow, R.W. and Crain, J.S., In *Applications of Inductively Coupled Plasma-Mass Spectrometry to Radionuclide Determinations, ASTM STP 1291*. R.W. Morrow and J.S. Crain, Eds., American Society for Testing and Materials, 1995.

[2] Alvarado, J.S. and Erickson, M.D., "Determination of Long-Lived Radioisotopes Using Electrothermal Vaporization - Inductively Coupled Plasma Mass Spectrometry," *Journal of Analytical Atomic Spectrometry*, Vol. 11, 1996, pp. 923-928.

[3] Chiappini, R., Taillade, JM. and Brébion, S., "Development of a High-Sensitivity Inductively Coupled Plasma- Mass Spectrometer for Actinide Measurement in the Femtogram Range," *Journal of Analytical Atomic Spectrometry*, Vol. 11, 1996, pp. 497-503.

[4] Garcia Alonso, J.I., Thoby-schultzendorff, D., Giovannone, B. and Koch, L., "Analysis of Long-Lived Radionuclides by ICP-MS," *Journal of Radioanalytical and Nuclear Chemistry, Articles*, Vol. 203, 1996, pp. 19-29.

[5] Hure, J., Rastoix, M. and Saint-James, R., "Données Relatives aux Equilibres Chimiques Régissant la Séparation Zirconium–Hafnium," *Analytica Chimica Acta*, Vol. 25, pp. 1-9.

[6] Tardy, P., "Caractérisation des Radioéléments Emetteurs Bêta à Vie Longue dans les Effluents de Retraitement des Combustibles : Application au Dosage à Très Bas Niveau du Palladium 107 et du Samarium 151," Thèse Université Aix-Marseille III, 1995.

[7] Shabana, R. and Ruf, H., "On the Use of Anion Exchange Resin and Acidic-Organic Mixtures in the Separation of Na, Cs and Ba," *Journal of Radioanalytical Chemistry*, Vol. 43, 1978, pp. 21-29.

[8] Noé, M., Révy, D. and Hérès, A., "Characterization of Critical Pure Long Life Beta Emitters for Radwaste Repository," *International Conference on Evaluation of Emerging Nuclear Fuel Cycle Systems*, Palais des Congrès, Versailles, 1995.

Bernard Mitterrand, [1] Pierre Leprovost,[1] Jacques Delaunay, [1] and Alain M. Vian [1]

DETERMINATION OF TECHNETIUM-99, NEPTUNIUM-237 AND ISOTOPES OF THORIUM IN URANYL NITRATE SOLUTIONS FROM A REPROCESSING PLANT, USING DOUBLE-FOCUSING ICP-MS

REFERENCE: Mitterrand, B., Leprovost, P. , Delaunay, J. and Vian, A.M., **"Determination of Technetium-99, Neptunium-237 and Isotopes of Thorium in Uranyl Nitrate Solutions from a Reprocessing Plant, Using Double-Focusing ICP-MS,"** *Applications of Inductively Coupled Plasma-Mass Spectrometry to Radionuclide Determinations*: *Second Volume, ASTM STP 1344*, R.W. Morrow and J.S. Crain, Eds., American Society for Testing and Materials, 1998.

ABSTRACT: The determination of some radionuclides in uranyl nitrate solutions from a reprocessing plant through chemical or radiochemical methods may be tedious, with poor precision . Quadrupole ICP-MS and, more recently, double-focusing ICP-MS, with high resolution capabilities, have proved to be very efficient tools for such determinations. These improvements will be illustrated by the examples of Technetium-99, Neptunium-237 and Thorium.

KEYWORDS: elemental mass spectrometry, ICP-MS, radiochemical methods, trace analysis, technetium, neptunium, thorium

---

Introduction

For six years Quadrupole ICP/MS (Q-ICP-MS) has been a powerful tool in our laboratory for assaying impurities in reprocessing products, uranyl nitrate [1] and plutonium dioxide. This technique brought a significant improvement relative to previous ICP - Optical Emission Spectrometry as elemental determinations could be performed without solvent removal of the heavy metal matrix. Because of ion combination interference, elements such as iron, calcium, and silicon were not determined very well. Phosphorus, sulfur and potassium could not be determined by Q-ICP -MS.
Using High Resolution ICP-MS (HR - ICP- MS) solved the problem [2].

[1] COGEMA Etablissement de La Hague Laboratory Division F 50444, Beaumont Hague Cedex, France

The double-focusing mass filter gives another advantage: the very low level of noise allows very low limits of determination. Thus HR-ICP-MS appears to be a potential challenger of conventional chemical or radiochemical techniques for the determination of long-lived radionuclides [3]. This paper gives examples of comparison for Technetium-99, Neptunium-237 and isotopes of Thorium.

Instrumentation

Instrument Description

Ion source -- The ion source is a horizontal plasma torch enclosed in a Faraday's box which is attached to a HF tuning box. The tuning box can be moved along the 3 directions (x, y, z) in order to perform alignment relative to the spectrometer interface. This device is in a glovebox. The spectrometer interface, holding the cones, is mounted on a plate, attached to a lateral wall of the glovebox (see Figure 1). The addition of the mounting plate is the only modification made to the instrument to connect it safely to the glovebox.

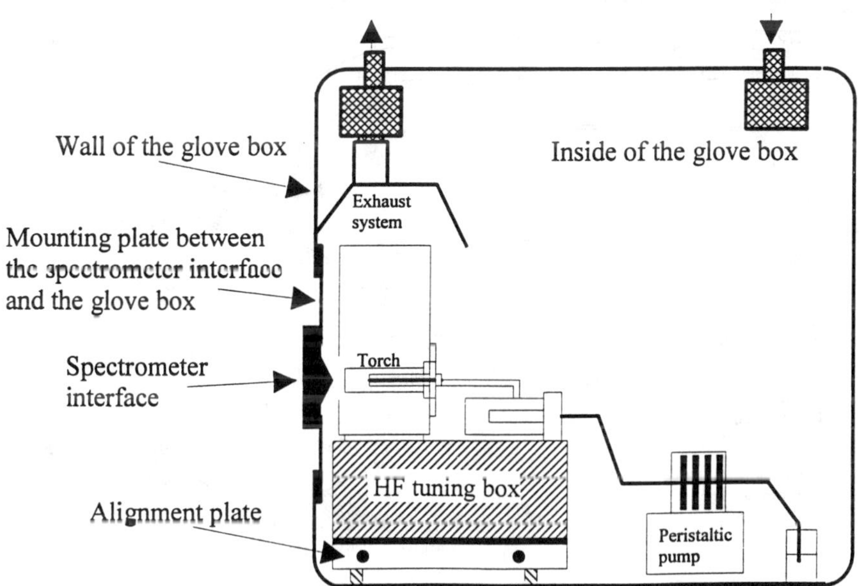

FIG. 1 -- *Schematic of the plasma torch in glovebox.*

Mass analyzer-- The mass analyzer is an "ELEMENT" spectrometer from Finnigan Mat®. The main parts are the following (see Fig. 2):

- a plasma interface with sampler and skimmer cones,
- a transfer and focusing ion optics,
- an acceleration and beam focusing device,
- a set of entrance slits,
- an electromagnet for mass filtering,
- an electric sector for energy filtering,
- a set of exit slits,
- a detection device with a dynode Secondary Electron Multiplier (SEM).

The set of slits allows the choice of the resolving power (10% of maximum height) among three values: 300, 3 000 and 7 500.

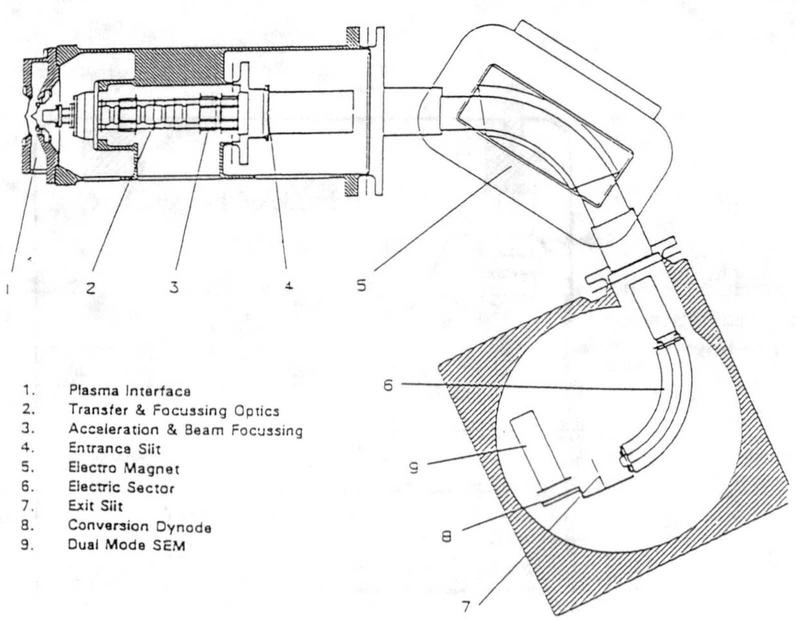

FIG. 2 -- *Schematic of the mass analyzer.*

Operating Conditions

Plasma torch -- The operating conditions for the plasma torch for all the measurements are listed in Table 1:

TABLE 1-- *Plasma torch operating conditions.*

Parameters	Values
Power	1 350 W
Reflected power	< 10 W
Plasma gas flow	12.5 1/ min.
Auxiliary gas flow	0.8 l/min.
Carrier gas flow	0.7 l/min.
Solution flow rate	1 ml/min.

Mass spectrometer -- The common operating conditions for the measurements of the different elements are listed in Table 2:

TABLE 2 -- *Mass spectrometer standard conditions.*

Parameters	Values
Lens compartment pressure	$< 10^{-4}$ mbar
Analyzer compartment pressure	$2 \cdot 10^{-8}$ mbar

Procedure

I. Chemical and Radiochemical Reference Procedures

Technetium-99-- Tc-99 is assayed as element Tc through a spectrophotometric analysis. The main steps of the method are the following [4]:

- reduction of Tc(VII) to Tc(V) with ascorbic acid,
- formation of Tc(V)-thiocyanate complex,
- extraction of the complex in butyl acetate,
- spectrophotometric measurement at 515 nm (ε 15 800), using known addition method.

Neptunium-237-- Np-237 is assayed through alpha spectrometry after solvent extraction separation. The oxidation states of Np in nitric acid media being IV, V and VI, a reductant such as ferrous sulphamate is used to produce Np(IV). This species is extracted with Thenoyltrifluoroacetone (0.1 M in xylene) from a 1 M nitric acid solution. The organic phase is measured by alpha spectrometry (Table 3). The extraction yield is monitored through a Np-239 spike, measured by gamma spectrometry.

Thorium-228-- Th-228 is assayed through alpha spectrometry after solvent extraction separation. Th(IV) is extracted with Thenoyltrifluoroacetone (0.2 M in xylene) from a nitric acid solution at pH 1. Thorium is back-extracted in 2 M nitric acid. The aqueous phase is measured by alpha spectrometry (Table 3).

TABLE 3--*Energy ranges for alpha spectrometry.*

Radionuclide	Energy
Np-237	4 725 keV
Th-228	5 340 keV
Th-228	5 423 keV

II. ICP/MS Procedure

Technetium-99--Standardization is performed by measuring a blank (0.5 M nitric acid) and a standard (Tc_0 = 10 μg/l in 0.5 M nitric acid) throughout the mass range 98.7 - 99.1 atomic mass units (a.m.u.). The resolution is set to 300. The result is the mean of 50 scans, 1 second of duration each. The sample, a reprocessing uranyl nitrate solution containing about 400 grams of uranium per liter, is diluted in 0.5 M nitric acid in order to get a solution containing $U_0 \approx 1$ g/l. Potential supression effects are corrected for using an indium internal standard ([In] = 10 μg/l in each solution). Further, limiting uranium concentration to 1 g/l avoids severe matrix effects ; the corresponding dilution is not too high, allowing a good limit of detection relative to the original sample (see Results and Discussion section). The measurements are performed with the same settings as for the standards.

Neptunium-237-- A solution of natural uranyl nitrate in 0.5 M nitric acid (U = 1 g/l) is spiked with 1 μg Np/l. This solution is measured throughout the mass range 236.5 - 237.5 a.m.u. (50 scans of 1 second each) ; the resolution is set successively to 300, 3000 and 7500. The ratio signal to background is measured in the three cases to determine the effect of resolution on detection of Np-237 (see Results and discussion section). Intensities are determined at the maximum of the peak ; the matrix matching allows correction for the slope of the baseline. A solution of unspiked reprocessing uranyl nitrate is analyzed according to the same measurement procedure.

Thorium -- Two analytical procedures were tested. In the first one, sodium fluoride is added to a sample of reprocessing uranyl nitrate solution; acidity is changed to 6 M by addition of concentrated nitric acid. Uranium is then extracted with Tributylphosphate (TBP), 30% in volume in dodecane. The aqueous phase is measured by ICP-MS in the range 227.0 - 233.0. In the second procedure, Th(IV) in another solution of reprocessing uranyl nitrate is extracted with Thenoyltrifluoroacetone (TTA), 0.2 M in xylene, from a nitric acid solution at pH 1. Thorium is back-extracted in 2 M nitric acid. The aqueous phase is measured throughout the mass range 227.0 -233.0.

Results and Discussion

Technetium-99-- Ten replicate measurements on a uranyl nitrate sample gave a mean peak intensity (N_b^{Tc}) of 171 counts per second (cps) with a standard deviation (σ^{Tc}) of 18 cps. The standard ($Tc_0 =$ 10 μg/l) produced a peak intensity of $2.09 \cdot 10^5$ cps (N_s^{Tc}); a uranyl nitrate sample, spiked with 10 μg Tc/l, gave $2.01 \cdot 10^5$ cps (N_x^{Tc}). The limit of detection (LOD^{Tc}) with 95 % certainty is computed according to the following statistical formula [5]

$$LOD^{Tc} = 4 \cdot [\sigma^{Tc} / (N_s^{Tc})] \cdot Tc_0 / U_0 = 0.004 \ \mu g \ Tc / g \ U$$

The recovery is 96% ; the concentration found in the spiked sample being

$$Tc_x = Tc_0 \cdot (N_x^{Tc}) / (N_s^{Tc}) = 9.6 \ \mu g/l$$

This high recovery shows that the measurement procedure with indium internal standard for standardization in nitric acid and sample uranium 1 g/l corrects well for suppression effects

Figure 3 shows the mass spectra within the Tc scanning range for a sample of reprocessing uranyl nitrate and a spiked sample (1 μg Tc/l). The high sensitivity coupled with a high signal-to-noise ratio explains the excellent limit of detection. The reference spectrophotometric method gives < 0.1 μg Tc / g U ; analysis time is 4 hours for a single monoelemental determination. In comparison, ICP-MS allows determination of Tc among 30 other elements in the same time, with better detection limit.

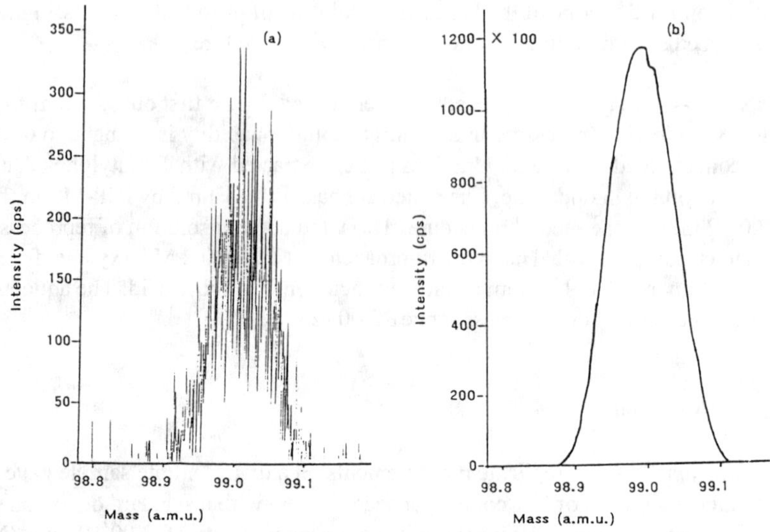

FIG. 3 -- *Mass spectra at mass 99 a.m.u. for a uranyl nitrate sample (a) and a spiked sample (1 µg/l) (b).*

Neptunium-237-- For Tc, the main component of background is electronic noise; the background for Np-237 is produced by a tail of the U-238 peak. Absolute sensitivity decreases as resolution increases but detection capability which is proportional to the signal-to-background ratio (S/B) increases dramatically (Table 4). Figure 4 shows the spectra with resolutions 300 (a) and 3000 (b) for the spiked sample. Figure 5 shows the spectra with resolution 7500 for the spiked sample (a) and the unspiked sample (b).

TABLE 4-- *Signal-to-background ratio for 1 ppm Np.*

Resolution	Signal (cps)	Background (cps)	S/B ratio
300	45 113	1 503 759	0.03
3000	15 526	10 921	1.42
7500	1 687	659	2.56

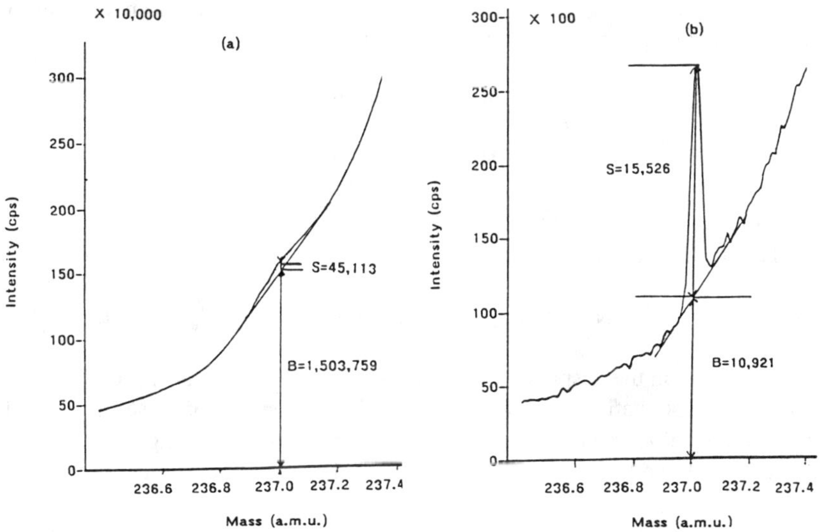

FIG. 4 -- *Mass spectra at mass 237 a.m.u. for Np 1 ppm Resolution 300 (a)- Resolution 3000 (b).*

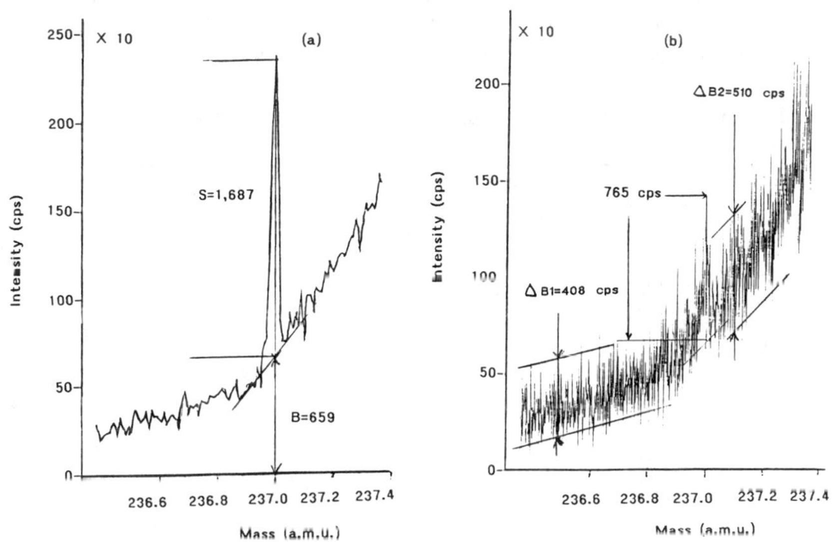

FIG. 5 -- *Mass spectra at mass 237 a.m.u. with resolution 7500 Np 1 ppm (a) - Unspiked uranyl nitrate (b).*

The standard deviation (σ_{Np}) of ICP - MS measurement is derived from a graphical evaluation of the range of the background (see Figure 5 (b))

$$\Delta B = (\Delta B_1 + \Delta B_2) / 2 = 459 \text{ cps}$$

$$\sigma^{Np} = \Delta B / 4 = 115 \text{ cps}$$

The instrumental response (N_s^{Np}) is 1 687 cps for a neptunium concentration (Np_0 / U_0) of 1 µg Np / g U. Thus, the detection limit for Np (LOD^{Np}) with 95% certainty is

$$LOD^{Np} = 4 \cdot [\sigma^{Np} / (N_s^{Np})] \cdot Np_0 / U_0 = \Delta B / (N_s^{Np})] \cdot Np_0 / U_0 = 0.27 \ \mu g \ Np / g \ U$$

The concentration in the unspiked sample (b) is 0.45 ± 0.14 µg Np / g U (0.14 is the confidence interval relative to random errors, mainly measurement noise: $2 \cdot \sigma^{Np}$); the determination by alpha spectrometry following the reference method gives 0.65 ± 0.07 µg Np / g U. The two figures are consistent, given the respective confidence intervals.

Thorium -- After extraction of uranium with TBP, isotopes 230 (half-life : $8 \cdot 10^4$ years) and 232 (half-life : $1.39 \cdot 10^{10}$ years), are easily detected on the mass spectrum (Figure 6); isotope 228 (half-life : 1.9 years) is seen, close to detection limit (Table 5). The peak at mass 231 is likely Protactinium 231 (half-life : $3.43 \cdot 10^4$ years), rather than Thorium 231; the half-life of Th-231 (26 hours) is too short to allow significant mass amounts to be present. The behaviour of Pa-231 under our conditions of extraction was not studied ; quantitation of this radionuclide with ICP-MS was not tried.

TABLE 5 -- *Thorium isotopes: atom % (measured after extraction of U with TBP).*

Isotope	Atom %
Th-228	< 1%
Th-230	31,3%
Th-232	68,7%

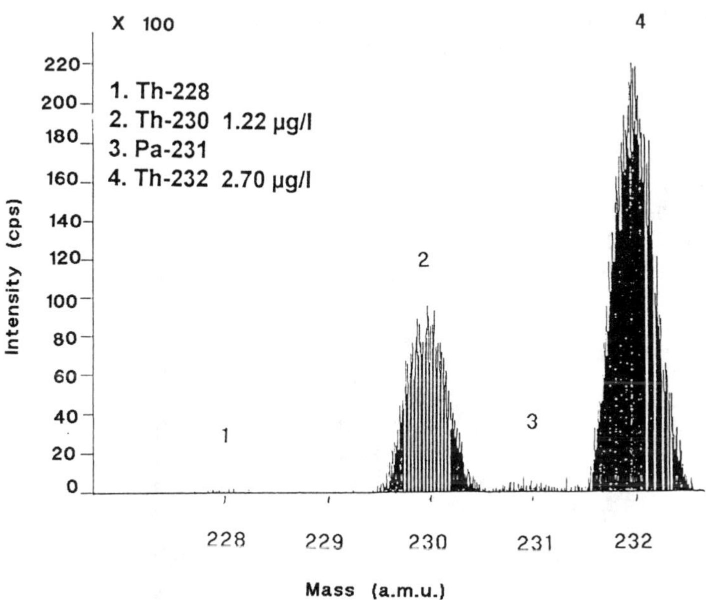

FIG. 6 -- *Mass spectrum of thorium after TBP extraction.*

The mass spectrum after extraction of thorium with TTA, and back-extraction in nitric acid is displayed on figure 7. Isotopes 230 and 232, are measured (Table 6). The preparation procedure is more complex than the TBP procedure, but has the advantage of concentrating thorium in the back-extraction stage. Thus Th-228 can be determined at an 8 ng/l level, while alpha spectrometry gives 7 ng/l. The interpretation of the peak at mass 231 has not been fully carried on ; nevertheless, Pa-231 is most likely. Quantitative extraction from hydrochloric solution can be performed with TTA [6] ; extraction yield from uranyl nitrate solutions has to be investigated more thoroughly.

TABLE 6 -- *Thorium isotopes: atom % (measured after extraction with TTA).*

Isotope	Atom %
Th-228	0,4%
Th-230	41,6%
Th-232	58,0%

The total counting time for ICP/MS is in the order of magnitude of a few minutes, while the figure is several hours for alpha spectrometry.

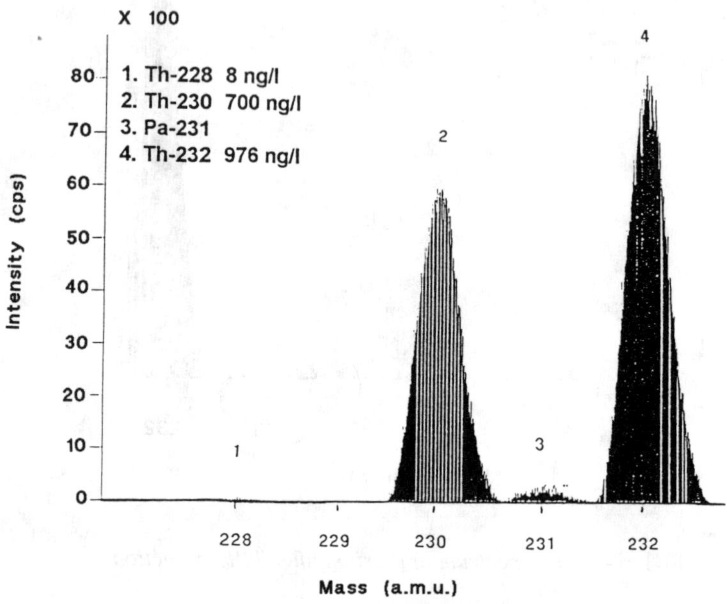

FIG. 7 -- *Mass spectrum of thorium after TTA extraction.*

Conclusion and Further Developments

The results show that HR-ICP-MS is able to determine Tc-99, Np-237 in uranyl nitrate solutions, in a more straightforward, rapid and efficient manner than chemical and radiochemical methods. Further, the use of organic solvents is avoided, giving a solution to the problem of organic waste management. This technique can be applied as well for analysis of plutonium nitrate solutions. The analysis of these heavy metal matrices requires a large dilution, driving to the need of high sensitivity. HR-ICP-MS matches easily the requirement for Tc-99 as well as enhanced Q-ICP-MS; but for Np-237, for which very high abundance sensitivity is also required, high resolution is indicated.

For thorium, solvent extraction cannot be avoided since the concentration of Th-228 is too low to be detected in presence of uranium. ICP-MS has nevertheless the advantage of giving results with shorter measurement times: a few minutes instead of hours for alpha spectrometry.

An extension to the isotopes of americium and curium would require prior separation of plutonium to avoid isobaric interference. Ion exchange separations are now widely available for this purpose. A more recent way, already in use in some laboratories, is coupling with liquid chromatography.

References

[1] Leprovost, P., Schott, R., and Vian, A., "Impurity Assay in Reprocessing Uranyl Nitrate", *Applications of Plasma Source Mass Spectrometry II*, The Royal Society of Chemistry, 1993 pp 213-221.

[2] Dall'Ava, D., Frémy, L., Nadisic, M., and Beloeil, J-Y., "La spectrométrie de masse à source plasma haute résolution - Son potentiel analytique", *Analyse Elémentaire par Spectrométrie de Masse*, CEA, CETAMA 1995.

[3] Provitina, O., Riglet, C., "Dosage de radionuclides à l'état de traces - Possibilités offertes par l'analyse en ICP/MS", *Rapport Technique* DSD N°59, CEA, 1992.

[4] Howard, D. H., and Webe,C.W., *Analytical Chemistry*, Volume 4, No. 4, April 1962, pp 530-533.

[5] Massart, D.,L., Vandeginste, B., G., M., Deming, S., N., Michotte, Y., and Kaufman,L., *Chemometrics: a textbook*, Elsevier 1988, pp 110-114.

[6] Stary, J., *The solvent Extraction of Metal Chelates*, Pergamon Press ltd. 1964, pp 75 and 191

Applications of ICP-MS in Health and Environmental Monitoring

E. Hywel Evans,[1] Jason B. Truscott,[1] Lee Bromley,[1] Phil Jones,[1] Justine Turner,[2] and Ben E. Fairman[2]

Evaluation of Chelation Preconcentration for the Determination of Actinide Elements by Flow Injection ICP-MS

REFERENCE: Evans, E. H., Truscott, J. B., Bromley, L., Jones, P., Turner, J., and Fairman, B. E., **"Evaluation of Chelation Preconcentration for the Determination of Actinide Elements by Flow Injection ICP-MS,"** *Applications of Inductively Coupled Plasma-Mass Spectrometry to Radionuclide Determinations: Second Volume, ASTM STP 1344*, R.W. Morrow and J.S. Crain, Eds., American Society for Testing and Materials, 1998.

ABSTRACT: A chelation column preconcentration method has been developed for the determination of uranium and thorium in waters by ICP-MS. Detection limits of 24 pg and 60 pg respectively were obtained, but these were blank limited. Uranium and Thorium were determined in certified reference materials. Results for uranium were 121 ± 21 and 15 ± 3 ng g^{-1} in NIST 1566a and NIST 1575 compared with certified values of 132 ± 12 and 20 ± 4 ng g^{-1} respectively. Results for thorium were 29 ± 8 and 28 ± 5 ng g^{-1} in NIST 1566a and NIST 1575 compared with indicative and certified values of 40 and 37 ± 3 ng g^{-1} respectively. The on-line separation of actinide radionuclides was achieved by selective elution of U, Th, Pu, Np, and Am.

KEYWORDS: ICP-MS, actinides, radioisotopes, chelation, chromatography

---

Inductively coupled plasma mass spectrometry (ICP-MS) can be used for the rapid determination of the concentration and isotopic composition of the actinide elements. The principal advantages of ICP-MS are speed and sensitivity, with the capability of determining all the radioisotopes within a minute, at concentrations as low as 1 picogram mL^{-1} (10^{-12} g mL^{-1}) in liquid samples. In addition, there is no need to separate the elements one from another, as there is in α-spectrometry, because this is achieved by the mass spectrometer, hence, the number of sample pre-treatment stages can be greatly reduced. However, it is still necessary to separate the radionuclides from the matrix, a procedure

---

[1] Senior lecturer, research student, research assistant and principal lecturer respectively, University of Plymouth, Dept. of Environmental Sciences, Drake Circus, Plymouth PL4 8AA, UK.

[2] Research technician and research manager respectively, LGC, Queens Road, Teddington, Middlesex TW11 0LY, UK.

for which column preconcentration methods are ideal. A number of resins have been used for the preconcentration and separation of the actinides. Recently a number of very specific chelating resins have become available which are particularly suited to this task. Some extraction procedures and application of these resins have been addressed by Horwitz [*1,2,3*]. Crain et al. [*4*] have quoted 20 fg mL^{-1} detection limits for ^{239}Pu and ^{235}U using TRU-Spec™ resin (Eichrom Industries Inc., Darien, IL) as a preconcentration step. Wyse and Fisher [*5*] reported a potential 3 fg detection sensitivity for plutonium using ICP-MS and TRU-Spec™ resin and concluded that results for ^{239}Pu in urine were comparable to those for alpha-spectrometry. Flow injection ICP-MS (FI-ICP-MS) has been used [*6*] with TRU-Spec™ resin and good results were obtained for ^{230}Th and ^{234}U for soil reference material TRM-4. Aldstadt et al. [*7*] also report good results for FI-ICP-MS using TRU-Spec™ resin for the determination of ^{238}U.

Experimental

Instrumentation

An inductively coupled plasma mass spectrometer (PlasmaQuad 2+, VG Elemental, Cheshire, UK) was used. Data were acquired using the time resolved analysis software, which allows time resolved monitoring of multiple isotopes, and manipulated off-line using MassLynx software (VG Elemental). Operating conditions are shown in Table 1. The flow injection manifold was interfaced with the ICP-MS instrument as shown in Fig. 1.

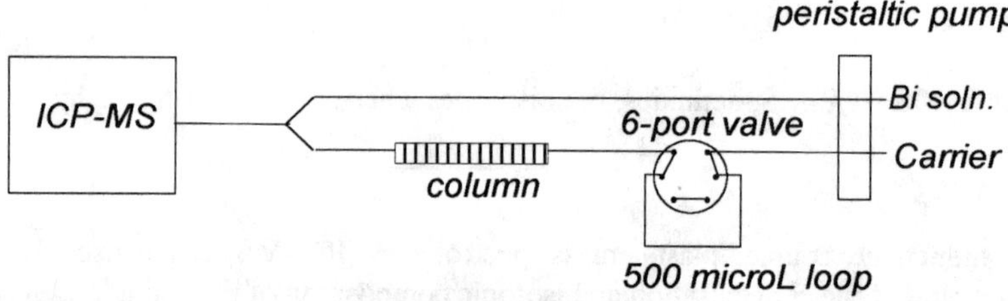

FIG. 1—*Schematic of the flow injection manifold interfaced with the ICP-MS.*

Determination of Natural Uranium and Thorium

Analytical Columns-- Columns were prepared, using a slurry of chelating resin (100-150 μm Tru-Spec resin, Eichrom Industries Inc., Darien, IL) in deionized distilled water (DDW), in commercially available glass chromatography columns of 3 mm i.d. and 50 mm length (Ominifit microbore columns, Omnifit, Cambridge, UK.

Reagents and Standards--Reagents were: 2 M HNO_3 (Analar, Fisher Scientific Ltd., UK); 5 ng mL^{-1} Bi internal standard; 1 M $Al(NO_3)_3$ (Analytical Grade, Fisher Scientific) purified by passing through a 10 x 0.5 cm column of Dowex 1-X8 anion exchange resin, then a 5 x 0.5 cm column of Tru-Spec resin; 0.5 M $Al(NO_3)_3$ + 2 M HNO_3 column feed solution; 0.1 M $NH_4HC_2O_4$ eluting solution.

A 1 µg mL^{-1} mixed standard solution of thorium and uranium was prepared in the column feed solution.

TABLE 1—*Operating conditions for ICP-MS.*

ICP	
Forward power (kW)	1.35 kW
Plasma gas (L min^{-1})	16.5
Auxillary gas (L min^{-1})	0.7
Nebulizer gas (L min^{-1})	0.8
Sampling depth (mm)	10
Sample flow (mL min^{-1})	0.5
Torch	Fassel (quartz)
Nebulizer	Concentric (quartz)
Spray Chamber	Scott type (quartz)
Interface	
Sampler	Ni, 1.0 mm orifice
Skimmer	Ni, 0.7 mm orifice
Pressure (mbar)	2 x 10^0
Mass Spectrometer	
Ion masses (m/z)	^{209}Bi, ^{230}Th, ^{232}Th, ^{235}U, ^{237}Np, ^{238}U, ^{239}Pu, ^{240}Pu, ^{242}Pu, ^{243}Am, ^{244}Pu
Data acquisition	Time resolved mode
Points per peak	3
DAC step	3
Dwell time (ms)	300
Time-slice duration (s)	1

Sample Preparation--The sample preparation procedure was partly based on a method by Nelson and Fairman [*8*]. Two certified reference materials (CRMs) were studied, namely NIST 1566a Oyster Tissue and NIST 1575 Pine Needles (National Institute of Science and Technology, Gaithersburg, USA). Samples (0.5 g) were dry ashed in crucibles at 200 °C for 2 hours, 400 °C for 2 hours, 600 °C for 2 hours, and 800 °C for 2 hours. This step was omitted for the oyster tissue. The samples were digested, with heating, in nitric acid (10 mL), boiled to dryness and heated on a hot hotplate. This was repeated until a white ash was left. On the last iteration the samples were boiled down until almost dry, and 10 mL of column feed soluion was added to dissolve the ash. Samples were made up to final volumes of 50 mL and 25 mL, for the oyster tissue and pine needles respectively, with column feed solution.

Calibration and Analysis—Calibrant and sample solutions were deposited onto the column, either by flow injection through a 500 μL loop or by uptake of a fixed volume, into a carrier stream of column feed solution at a flow rate of 0.5 mL min^{-1}. During deposition the outlet from the column was diverted to waste to prevent the column feed solution entering the ICP-MS instrument. After a deposition, the column was rinsed with 1 mL of 2 M HNO_3 to remove any residual column feed solution before the column was diverted back to the ICP-MS. The analytes were eluted with 0.1 M ammonium bioxalate, and the analyte masses monitored. Typical elution profiles for uranium and thorium are shown in Fig. 2. Peak areas were determined and ratioed to the continuous signal for Bi to compensate for instrumental drift.

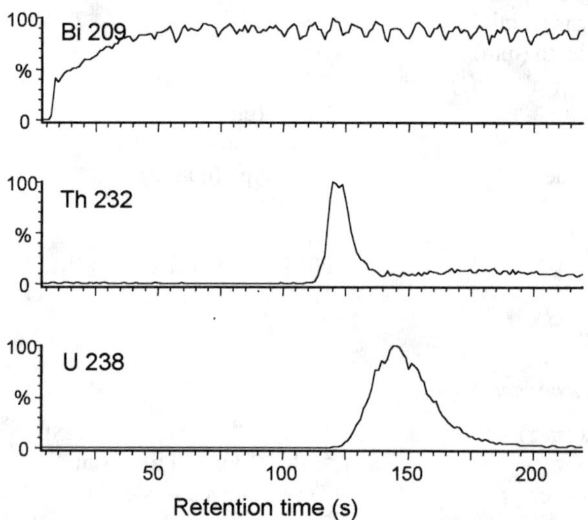

FIG. 2—*Elution profiles for 50 pg depositions of U and Th monitored at m/z 238 and 232 respectively.*

It is evident from Fig. 2 that uranium eluted completely over a period of 70 s and thorium over 30 s. In the case of thorium there was a raised backgound after elution of the peak, which may be due to incomplete elution from the column, impurities in the eluting solution, or tailing caused by the spray chamber. The elution times corresponded to volumes of approximately 0.6 and 0.25 mL for uranium and thorium respectively. Absolute detection limits were 24 pg and 60 pg for uranium and thorium respectively for a 500 μL sample, and were blank limited. Relative detection limits can be improved by greater preconcentration factors if the reagents are cleaned more thoroughly.

Determination of Actinide Radioisotopes

Analytical Columns--Low pressure preparative chromatography columns, 10 cm long and 0.5 cm i.d. (Econo-column, Bio-rad) were used to preconcentrate and/or separate Th, U, Pu, Am, and Np.

Reagents and Standards--Reagents were prepared as previously. A 10 ng mL^{-1} standard stock solution of each of the isotopes ^{230}Th, ^{232}Th, ^{235}U, ^{237}Np, ^{239}Pu, ^{240}Pu, ^{242}Pu, ^{243}Am, and ^{244}Pu was used for spiking experiments.

Sample Preparation—Samples (50g) were weighed into evaporating basins, placed in a muffle furnace and dry-ashed at 200 °C for 2 hours, 400 °C for 2 hours and 600 °C for 2 hours. Concentrated acid (50 mL) was added to each, left to stand for one hour, heated gently on a hot-plate until all nitrous oxide fumes were driven off, then boiled for ten minutes and allowed to cool. The samples were centrifuged for five minutes at 3500 rpm, the supernatant decanted and the sediment was re-extracted as before. The combined supernatant was boiled down until precipitation just began to occur, when an equal amount of the column feed solution was added to re-dissolve the precipitate. A portion (0.5 mL) of reducing solution, comprising 3 g of iron ammonium sulphate and 3 g of sodium formaldehyde sulfoxylate dissolved in 10 mL of 2 M HNO_3, was added to each sample and allowed to stand for 15 min. This ensured any iron present was reduced to Fe(II) to avoid column interferences.

Analysis of Samples— Samples were transferred to the columns, and the beakers were rinsed with 5 mL of column feed solution which was also added to the columns. They were then washed with two 5 mL portions of 1 M HNO_3 to ensure that no $Al(NO_3)_3$ remained, eluted with approximately 10 mL of 0.1M ammonium bioxalate and made up to volume. External calibration was performed and the samples analysed.

On-line separation—In this case samples were loaded onto the column as before, but eluted sequentially using an elution method developed by Horowitz et al. [*1*]. The method is represented schematically in Fig. 3.

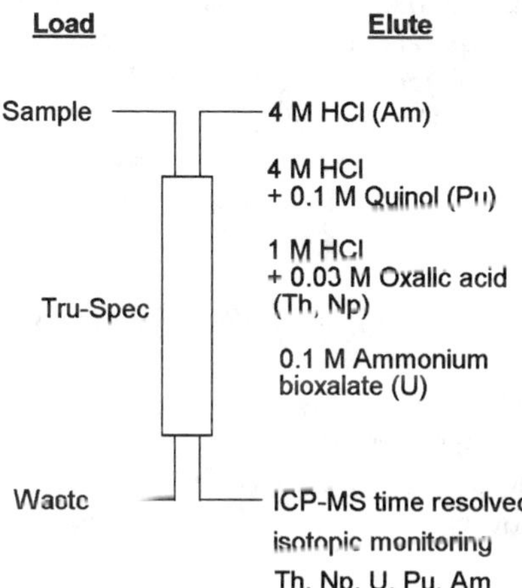

FIG. 3--*Schematic of elution method used for on-line separation of actinide radioisotopes.*

Results and Discussion

Determination of Uranium and Thorium

Results for the determination of uranium and thorium in oyster tissue and pine needles are shown in Table 2.

TABLE 2--*Results of the determination of uranium and thorium in certified reference materials by ICP-MS.*

CRM	No. of samples	U		Th	
		Certified value (ng g^{-1})	Found[a] (ng g^{-1})	Certified value (ng g^{-1})	Found[a] (ng g^{-1})
NIST 1566a oyster tissue	5	132 ± 12	121 ± 21	(40)[b]	29 ± 8
NIST 1575 pine needles	11	20 ± 4	15 ± 3	37 ± 3	28 ± 5

[a] $\bar{x} \pm s$.

[b] Uncertified indicative value.

For oyster tissue there was no significant difference between the found value and the certified mean for uranium at the P = 0.05 level. For the pine needles, low recoveries for both thorium and uranium were observed compared with the certified mean, though there was no significant difference between the found value and the bottom of the certified range for uranium (i.e. 16 ng g^{-1}) at the P = 0.05 level. This may indicate that uranium was associated with silicaceous material present in this sample, or that losses occurred during the ashing stage due to adsorption onto the surface of porcelain crucibles at high temperature [9]. Likewise, low recoveries for Th might have been due to association with silicaceous material, however, Th has a tendency to adsorb onto glassware, pump tubing, and the column resin, often resulting in low recoveries or high blank values. Ammonium bioxalate should be a sufficiently is a strong chelating agent to prevent such memory effects, however, other chelating agents might improve recoveries.

Recovery of Actinide Radioisotopes from Biological Samples

Samples of NIST 1566a oyster tissue were spiked with a solution of mixed actinide elements (10 ng g^{-1}), subjected to the dissolution and column preconcentration method, and recoveries determined by external calibration ICP-MS. Details of the samples are given in Table 3. Recovery data for the spiked samples after blank subtraction are shown in Table 4.

Recoveries for the spiked oyster tissue samples were within the range 88-107% (i.e. sample 2). However, recoveries of between 73-90% were obtained for the control sample (i.e. sample 3). The results for sample 3 were probably low because there was no sample matrix present to prevent absorption onto the walls of the beakers.

TABLE 3--*Details of spiked samples used for recovery tests.*

Sample No.	Sample type	Mass of oyster tissue (g)	Mass of spike (g)
1	Unspiked sample	0.5504	
2	Spiked sample	0.5930	2.0870
3	Control (no matrix)		2.0854
4	Blank		

TABLE 4--*Recovery of actinide radioisotopes from oyster tissue.*

	Conc. Of isotope in fraction (ng mL^{-1})								
	^{230}Th	^{232}Th	^{235}U	^{237}Np	^{239}Pu	^{240}Pu	^{242}Pu	^{243}Am	^{244}Pu
Sample 2									
Expected conc. (ng mL^{-1})	0.84	0.84	0.84	0.84	0.84	0.84	0.84	0.84	0.84
Actual conc. (ng mL^{-1})	0.90	0.80	0.74	0.79	0.77	0.77	0.79	0.84	0.77
Recovery (%)	107	95	88	94	92	92	94	100	92
Sample 3									
Expected conc. (ng mL^{-1})	0.84	0.84	0.84	0.84	0.84	0.84	0.84	0.84	0.84
Actual conc. (ng mL^{-1})	0.76	0.71	0.62	0.61	0.62	0.61	0.63	0.72	0.67
Recovery (%)	90	84	74	73	74	73	75	86	80

Recovery of Actinide Radioisotopes from Sediment Samples

Several different leaching and column methods were attempted for 50 g of spiked sediment. The sediment was chosen to present a worse case scenario. That is, it was a sieved and homogenized sediment which had been collected from the river Tamar in the Southwest of England, and contained very high levels of uranium, thorium, lanthanides, and iron, all of which act as interferents during column preconcentration. The results of nitric acid and nitric/hydrochloric acid leaches are given in Tables 5 and 6 respectively.

TABLE 5--*Recovery of actinide radioisotopes from sediment using a nitric acid leach.*

	Conc. of isotope in fraction (pg mL^{-1})								
	^{230}Th	^{232}Th	^{235}U	^{237}Np	^{239}Pu	^{240}Pu	^{242}Pu	^{243}Am	^{244}Pu
Expected conc. (pg mL^{-1})	200	200	200	200	200	200	200	200	200
Actual conc. (pg mL^{-1})	105	nd	Nd	104	144	153	143	<1	144
Recovery (%)	52	nd	Nd	52	72	76	72	<1	72

TABLE 6--*Recovery of actinide radioisotopes from sediment using a nitric/hydrochloric acid leach.*

	Conc. of isotope in fraction (pg mL^{-1})								
	^{230}Th	^{232}Th	^{235}U	^{237}Np	^{239}Pu	^{240}Pu	^{242}Pu	^{243}Am	^{244}Pu
Expected conc. (pg mL^{-1})	200	200	200	200	200	200	200	200	200
Actual conc. (pg mL^{-1})	87	nd	nd	84	136	132	130	<1	122
Recovery (%)	44	nd	nd	42	68	66	65	<1	61

Recoveries obtained using the two acid leaches were between 42-76% for most of the radioisotopes, with the exception of Am, which is the least well retained element. The low recoveries were probably due to column overloading by Fe, the lanthanides, and naturally occurring uranium and thorium, which were present in this sediment in excess, thereby causing column overloading. To obtain full recoveries it will be necessary to develop chromatographic methods to separate the actinides from other species, with sufficient resolution. We are currently investigating the use of new substrates with immobilized chelating dyes to achieve this aim.

On-line Separation of Actinide Radioisotopes

One of the problems associated with the determination of the actinide elements by ICP-MS is the propensity for polyatomic ion interferences. A particular problem is the determination of ^{239}Pu in samples containing an excess of naturally occurring uranium, due to the positive interference caused by ^{238}UH. One way to overcome this is the separation of the elements prior to ICP-MS detection. This was achieved by sequential elution of the actinides as shown in Fig. 4. In this case Am and Pu were separated from Np, U and Th, thereby eliminating the interference of ^{238}UH on ^{239}Pu. Complete separation of Am and Pu was not obtained, probably due to the propensity for Pu to disproportionate and exist in a number of oxidation states simultaneously.

Chromatographic methods can also be used to reduce problems of column overloading, provided that sufficient resolution is achieved. Such on-line chromatographic methods using novel resins and chelating dyes are currently being investigated in our laboratory.

Acknowledgements

The work described in this paper was supported in part by the Department of Trade and Industry, UK, as part of the Government Chemist Programme.

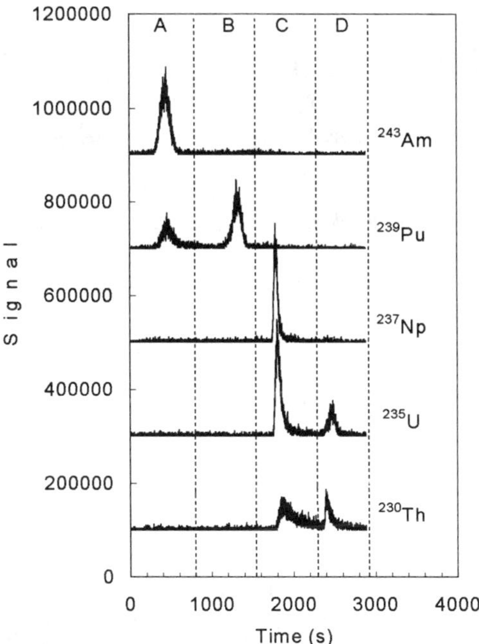

FIG. 4--*Elution profile for sediment spiked with 20 ng each of the actinide elements. A, 4 M HCl; B, 4 M HCl + 0.1 M quinol; C, 4 M HCl + 0.03 M oxalic acid; D, 0.1 M ammonium bioxalate.*

References

[*1*] Horwitz, E.P., Dietz, M.L., Chriarizia, R., Diamond, H.,Essling, A.M., Graczyk, D., *Anal. Chim. Acta*, 1992, **266**, 25.

[*2*] Horwitz, E.P., Chiarizia, R., Dietz, M.L., Diamond, H., *Anal. Chim. Acta*, 1993, **281**, 361,

[*3*] Horwitz, E.P., Dietz, M.L., Chriarizia, R., Diamond, H., Maxwell, S.L., Nelson, M.R., *Anal. Chim. Acta*, 1995, **310**, 63.

[*4*] Crain, J.S., Smith, L.L., Yaeger, J.S., Alvarado, J.A., *J. Radioanal. Nucl. Chem., Articles*, 1995, **1**, 133.

[*5*] Wyse, E.J., Fisher, D.R., *Radiation Protection Dosimetry*, 1994, **55**, 199.

[*6*] Hollenbach, M., Grohs, J., Mamich, S., Marilyn, K., *J. Anal. At. Spectrom.*, 1994, **9**, 927.

[7] Aldstadt, J.H., Kuo, J.M., Smith, L.L., Erickson, M.D., *Anal. Chim. Acta*, 1996, **319**, 135.

[8] Nelson, D.M., and Fairman, W.D., paper presented at the 36th Annual Conference on Bioassay, Analytical and Environmental Radiochemistry, Oak Ridge, Tennessee, 1990.

[9] Neumann, W.F., Fleming, R.W., Carlson, A.B., Glover, N., *J. Bio. Chem.*, 1948, **173**, 41.

Zeev Karpas,[1] Avraham Lorber,[1] Ludwik Halicz,[2] and Itai Gavrieli[2]

Determination of Uranium in Clinical and Environmental Samples by FIAS-ICPMS

Reference: Karpas, Z., Lorber, A., Halicz, L., and Gavrieli, I., "**Determination of Uranium in Clinical and Environmental Samples by FIAS-ICPMS**", *Applications of Inductively Coupled Plasma Mass Spectrometry to Radionuclide Determinations: Second Volume, ASTM STP 1344*, R. W. Morrow and J. S. Crain, Eds., American Society for Testing and Materials, West Conshohocken, PA, 1998.

Abstract: Uranium may enter the human body through ingestion or inhalation. Ingestion of uranium compounds through the diet, mainly drinking water, is a common occurrence, as these compounds are present in the biosphere. Inhalation of uranium-containing particles is mainly an occupational safety problem, but may also take place in areas where uranium compounds are abundant. The uranium concentration in urine samples may serve as an indication of the total uranium body content. A method based on flow injection and inductively coupled plasma mass spectrometry (FIAS-ICPMS) was found to be most suitable for determination of uranium in clinical samples (urine and serum), environmental samples (seawater, wells and carbonate rocks) and in liquids consumed by humans (drinking water and commercial beverages). Some examples of the application of the FIAS-ICPMS method are reviewed and presented here.

Keywords: bioanalytical, flow injection analysis, ICPMS, industrial hygiene

---

Uranium compounds are more commonly abundant in the biosphere than generally recognized and may be taken up by humans through two main pathways: ingestion and inhalation. Ingestion of uranium compounds may occur through human diet, mainly drinking water, while inhalation of dust particles containing uranium may

---

[1]Senior scientist and head, respectively, Department of Analytical Chemistry, Nuclear Research Center, Negev, Beer-Sheva, Israel.

[2]Senior scientists, Geochemistry Department, Geological Survey of Israel, Jerusalem, Israel.

take place particularly in areas where uranium compounds are abundant. A major part of the ingested uranium is excreted in feces. Only about 1% of the ingested uranium does cross the intestine-blood barrier, i.e., gets taken up, enters the bloodstream and later is mostly removed by the kidneys and excreted in urine. Inhaled uranium may be expectorated from the lungs through coughing or in saliva, while part of the uranium may be rapidly dissolved by lung-fluids and excreted in urine, and the rest may either be slowly dissolved or remain in the lungs for an extended period. This uranium absorbed by either pathway enters the bloodstream and may be removed by the kidneys and excreted in urine or be taken up by different organs, especially the bones and liver. In order to estimate the health hazards posed by uranium compounds, and to study the fate of uranium taken up by humans, analytical procedures have to be developed. These procedures must be sensitive, inexpensive, free of interferences and suitable for application in a variety of environmental and clinical matrices. The method of choice is Flow Injection Analysis Spectrometry (FIAS) coupled to an Inductively Coupled Plasma Mass Spectrometer (ICPMS). Some facets of this method and its application to clinical samples, water from different sources and beverages, are reviewed and described in the present work. The uranium content in drinking water may be indicative of its source: local wells, a local or national reservoir, or spring water. The uranium content in drillings or wells may help identify the aquifer, or body of water, upon which these are situated.

Experimental Method

Instrumentation

A commercial ICPMS instrument, ELAN-6000 by Perkin-Elmer/Sciex (Thornhill, Ontario, Canada) located at the Geological Survey of Israel (GSI) in Jerusalem, was used. The instrument, described in detail previously by Denoyer [*1*], was equipped with a standard plasma torch and a flow-injection system (FIAS 400 by Perkin-Elmer, Uberlingen, Germany). Instrument operation is fully controlled by a computer through a Windows NT driven dedicated software package, which also serves to process the data and calculate the results. The operational parameters of the instrument were optimized for detection of uranium, i.e., transmission of heavy ions was favored, as detailed previously [*2*].

Flow Injection System - The flow injection system was described in detail recently [*2*]. It was configured with two inputs: the carrier stream, which consisted of 0.6% nitric acid and 1 g/L Triton-X 100, and the sample stream, and two outputs: to the ICPMS and to the drain, and also included an injection loop (100 μL volume). All streams were controlled by a five-port valve, configured to operate either to load the loop or to inject the sample [*2*]. The analysis cycle starts with switching the injection valve from *loop loading* to *injection* for six seconds, which is sufficient time to carry the sample outside the injection head. This makes it possible to start cleaning the loop and filling it with the next sample while the analysis is taking place. Two rinse solutions are used to clean the loop and tubing for 14 seconds and in the remaining time of the cycle the next sample flushes the loop and fills it. This rinsing period is sufficient for samples with low uranium content (below 0.1 ppb) which may be measured in rapid succession, so that the total analysis time is about 60 seconds. However, additional rinsing may be required after samples with high levels of uranium in order to attain a low baseline level for the next measurement.

Typically, the intensity of the $^{238}U^{+}$ signal is sampled for 250 ms, so that each second 4 data points are acquired. This dwell time is a compromise between the long integration times desired for stable signals and the short dwell time needed for full coverage of the transient nature of the FIAS signal. The dwell time is adjusted according to the number of ions monitored, so that if internal standards are used, it may be shorter so that each ion is sampled at least twice per second. Averaging the readings acquired during the first sixteen seconds (before the sample reaches the ICP) yields the background level which is subtracted from the forthcoming transient flow-injection signal. Typical ICPMS traces of the intensity of the $^{238}U^{+}$ signal in the blank solution and two urine samples containing 0.006 and 0.020 ppb, are shown in Figure 1. The total analysis time is less than 60 seconds, and the section used for the baseline subtraction is shown.

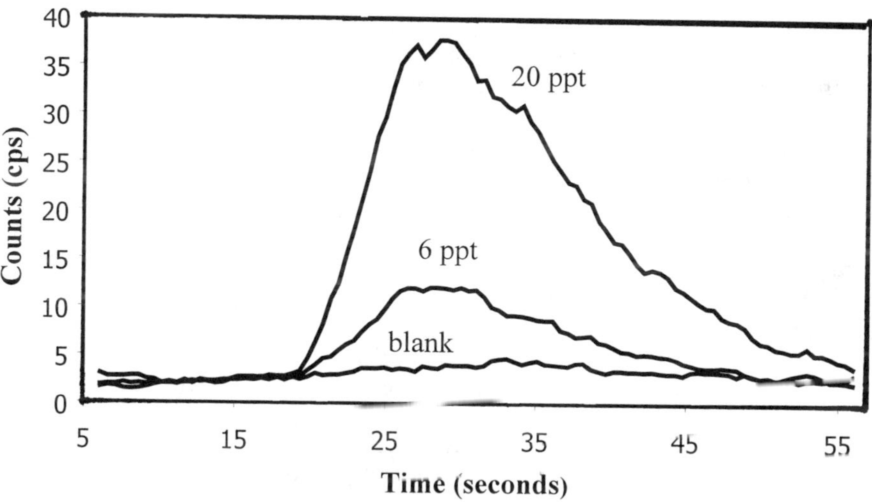

Figure 1 - Typical ICPMS traces of the intensity of the $^{238}U^{+}$ signal in the blank solution and two urine samples containing 0.006 and 0.020 ppb.

Sample Preparation

Clinical samples (urine and serum) required no sample preparation and were injected directly from the FIAS loop into the plasma torch of the ICPMS. Ocean water, drinking water and beverages were also measured without sample pretreatment. Water samples from environmental sources were measured after filtration to remove particulate matter and acidification. The high salinity Dead Sea samples were diluted and measured by flow injection analysis. Carbonate rock samples were treated in dilute nitric acid after finely grinding the rock samples.

Carrier and rinsing solutions -- The carrier and rinsing solutions were prepared from purified water (Nanopure by Barnstead, Dubuque, Iowa). Analysis by the ICPMS showed that the level of uranium in the purified water was below 0.0001 ppb. A 0.6% nitric acid solution was prepared (65% Suprapur grade, Merck, Darmstadt, Germany) and 1 g/L Triton-X 100 (also from Merck) was added to the solution. Measurements by the ICPMS showed that the level of uranium in the carrier solution was about 0.0001 ppb.

Calibration -- For aqueous samples the instrument was calibrated with a standard solution (ICP-VI Multielement standard, Merck) containing 10 ppb of uranium. For urine samples a calibration solution with 1 ppb of uranium was prepared by adding uranyl nitrate (Merck) to a composite urine sample taken from several "unexposed" individuals. This calibration procedure somewhat corrects for the matrix effects arising from the urine samples. In measurement of highly saline solutions, such as seawater samples, a solution containing 10 ppb of thallium was used as an internal standard. Although other ions with masses above 200 daltons may also be used, thorium [*3*] and bismuth [*4*] were reported as introducing memory effects, and thallium was found to be satisfactory as an internal standard.

Results and Discussion

1. Clinical Samples

As demonstrated recently [*2, 5-8*], and shown in Figure 1, FIAS-ICPMS is particularly well suited for determination of uranium in urine. The sensitivity of the method, which has a detection limit of 0.001 ppb, is sufficient for measurement of background levels of uranium in urine for the normal population (people who are not occupationally exposed to uranium compounds), which has a mean of about 0.012 ppb [*6*]. It was also shown previously that the concentration of uranium in urine, measured in "spot samples", may vary considerably from voiding to voiding of the same individual by a large factor (in one case, a factor of five was found within the span of a few hours) [*7*]. This can be overcome in part by normalizing the uranium concentration to the creatinine concentration in the same spot sample, and expressing the results in units of [ng Uranium /g Creatinine]. As the diurnal creatinine excretion for each individual may be calculated with reasonable confidence from the weight, height and age of the subject [*9*], this may serve as a convenient way for calculating the body-burden of uranium from a "spot sample" of urine and is useful for internal dosimetry purposes. It should be noted that the urine sample can be collected directly in a sterile test-tube, which does not contribute any measurable amount of uranium to the background level.

Like urine samples, serum samples do not require pretreatment prior to analysis. However, serum samples are collected by an intrusive procedure: a blood sample must be drawn by a syringe needle into a test-tube and then the serum must be separated from whole blood prior to analysis. These procedures could introduce some cross-contamination and introduce an artifact into the analytical results, which may be the reason for discrepancies in the reported "normal" values, which according to recent studies may be below 0.001 ppb [*10*].

2. Liquid Samples

2.1 Seawater - Flow injection-ICPMS was previously used to determine the content of uranium in sea water [*2*] and the same method was used in the present work. A NASS-4 ocean water standard, with a certified uranium concentration of 2.68±0.12 ppb, was found to contain 1.9-2.1 ppb, i.e., a suppression of 22-30% due to the high salt content in the sample. However, using a solution containing 10 ppb of thallium as an internal standard compensated for this matrix effect, and a value of 2.72±0.08 ppb was found. Determining the uranium content in water from the Dead Sea presents a particularly tough problem due to the extremely high content of total dissolved solids (337 g/L and a density of 1.235 g/mL). Halicz et al. [*11*] used a silicone immobilized 8-hydoquinoline column to separate and preconcentrate uranium on-line and then measure it by ICPMS, and obtained a concentration of 2 ppb. A result of 1.75±0.06 ppb was obtained in this study by using the method described above, FIAS-ICPMS with thallium as an internal standard, for Dead Sea samples diluted 20-fold and 40-fold. This may seem rather low compared to ocean water (see above), considering that the salt content in the ocean is an order of magnitude lower than that of the Dead Sea.

2.2 Springs and Wells in Arid Areas - The 39 active wells that were surveyed supply water to the town of Eilat and to the settlements in the Arava Valley (connecting the Dead Sea with the Red Sea) in the south of Israel (Figure 2). Water that is pumped at these drillings originates from several aquifers that lie in different geological layers with depths ranging from 39 to 960 m. Some of these are interconnected and some are separated from other subterranean bodies of water. The hydrogeology of this system is complex, and it is not always easy to correlate a given well with a unique aquifer, a factor of importance for monitoring the migration of pollutants or for assuring water quality. The content of several trace elements may serve as a means of identifying wells that draw upon the same aquifer. It was found that wells that are geographically close to one another are not necessarily situated on the same aquifer, and this may not be reflected by the salinity (or chloride concentration). The uranium concentration, that was found to vary from 0.15 μg/L to 15 μg/L among these 39 wells, may serve as such a tracer. For example, in Yaalon 116 (see Figure 2) the uranium content was found to be 8.6 ppb (Cl^- content was 400 ppm), in nearby Yaalon 8 a level of 2.8 ppb (Cl^- was 700 ppm) was found, while the concentration of uranium in Yaalon 1, 3 and 7 was 0.6, 1.8 and 0.2 ppb (Cl^- levels were 430, 500 and 400 ppm), respectively. Indeed the former is situated on the Cenomanian aquifer, Yaalon 8 is on the Hazeva aquifer while the latter drillings are located on the Lower Cretaceous aquifer. Like the chloride content, most other elements and the total dissolved solids do not differ this significantly.

2.3 Drinking Water in Israel and in Beverages - The uranium content in drinking water in Israel reflects the source from which the water is obtained. The Sea of Galilee (Lake Kinneret as it is known locally) is the major source of drinking water, and supplies most of Israel's needs through a cross-country water conduit (the National Water Carrier). The Banias and Dan springs, which are two of the major sources of the Jordan River that feeds this lake, have low uranium contents (see Figure 2). The average uranium concentration, from several samples taken at different locations and depths, in the water of the Sea of Galilee, is around 0.58 ppb. Slightly higher concentrations of uranium were found that in tapwater in regions supplied almost solely from this source, even if they are far away like Dimona and Yeroham (about 300 km away), or in Jerusalem (see Figure 2). In some areas, local wells and drillings also contribute to the drinking water supply. In Beer-Sheva, for example, the uranium content in tap water is around 5 ppb due to the contribution of the local wells, while in nearby settlements it ranges from 3 ppb in Omer (5 km NE) to 1.5 ppb further away in Metar (13 km NE) and Arad (50 km E) as shown in Figure 2.

TABLE 1--*Uranium content in Israeli and other beverages.*

Sample	U, ppb	Sample	U, ppb
Cola	0.15, 0.03	Mineral (Israel-M)	0.75, 0.44, <0.01
Cola (diet)	0.02	Mineral (Israel-E)	1.4
Soda water	0.08, 0.21	Mineral (Israel-N)	0.50, 0.44
Beer (Israeli)	0.03	Mineral (Belgian)	<0.01
Apple Drink (diet)	0.27	Mineral (Romania)	<0.01
Before Filter (B)	0.6	Mineral (French)	2.0
After Filter (B)	0.02	Orange Drink (natural)	0.03
		Grapefruit Drink (diet)	0.30

On the whole, beverages sold by internationally renowned commercial firms, are produced according to strict quality control practices, which involve pretreatment of the water, usually on some ion-exchange column. Thus, the concentration of uranium in these beverages (Table 1) is well below that of the local tap-water, as shown in Figure 2.

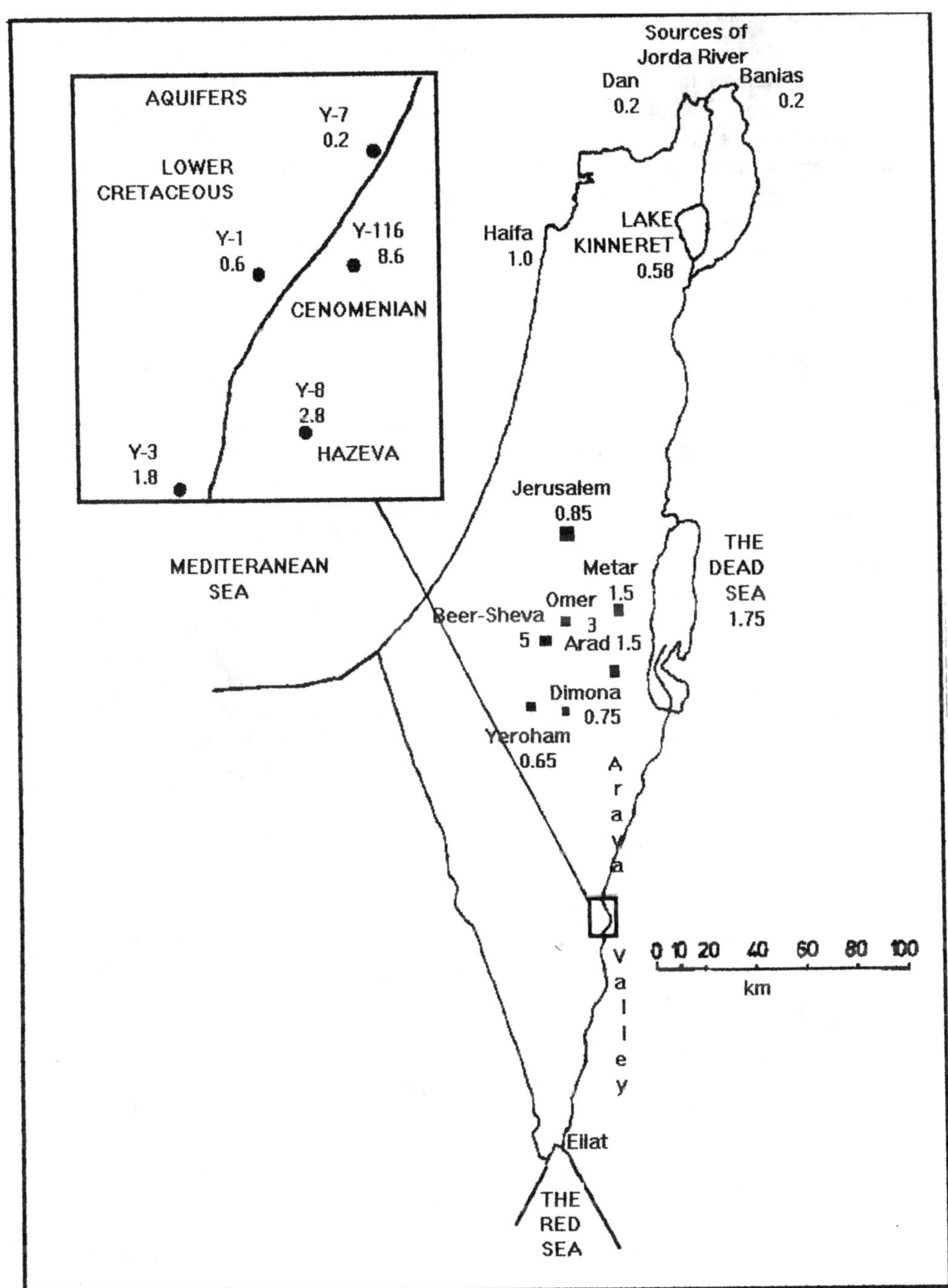

Figue 2 - A Map of Israel Showing the Locations and Uranium Concentrations, in ppb, Discussed in the Text. INSET Shows the Yaalon Drillings in the Arava Valley

However, it should be noted that measurement of a variety of samples from the same firm may yield quite different values, depending on the batch sampled. Beverages that are sold as "mineral water" or "natural spring water" may contain uranium in concentrations that vary by several orders of magnitude. Some brands, with high mineral content (and very high uranium levels in the 10-100 ppb range), even boast their therapeutic properties. On the other hand, some brands of mineral water have such low uranium concentrations (below 0.01 ppb) that one may wonder if some minerals were added to deionized water, rather than originating from a natural source.

3. Carbonate Rocks

Carbonate rocks, which are abundant in the Earth's crust, are composed mainly of calcium and magnesium carbonate, and may also contain uranium [*12*]. In fact, the ratio between uranium and its progeny from radioactive decay, thorium, the $^{230}Th/^{234}U$ ratio, may serve as a means of dating the rocks. The method of choice for the accurate measurement of this ratio is based on Thermal Ionization Mass Spectrometry (TIMS). A rapid method for screening the samples and selecting those with high uranium content is desirable, as sample preparation for TIMS is tedious. A method based on dissolution of finely ground rock samples in dilute nitric acid, and determination of the uranium and thorium content by FIAS-ICPMS with thallium as an internal standard, proved to be most suitable, as described previously [*12*]. As shown in Figure 3, the results compared favorably with measurements by Thermal Ionization Mass Spectrometry (TIMS), where the uranium content in the rock samples was in the range of 250-600 ppb.

Summary

The use of a flow injection system for introducing samples into the ICPMS may find several applications in determining the level of uranium in environmental and clinical samples. The method is rapid (1 minute per measurement), sensitive (limit of detection 0.001 ppb) and determination of uranium is free of isobaric interferences. The fact that over 40 samples an hour, in addition to blanks and standards, may be measured by a single technician also make this a cost-effective and inexpensive method (estimated as less than $10 per sample). Memory effects between samples may be considerably reduced by use of a surfactant (Triton 100-X) and by rinsing the tubing and sample loop during the measurement of the previous sample. This has been demonstrated for clinical samples and leads to a better estimate of uranium body-burden for internal dosimetry purposes, particularly after normalization to creatinine levels in urine spot samples [*7*]. One of the major sources for ingested uranium is in the beverages consumed by individuals, and it was shown here that these vary considerably and depend on the dietary habits of each subject.

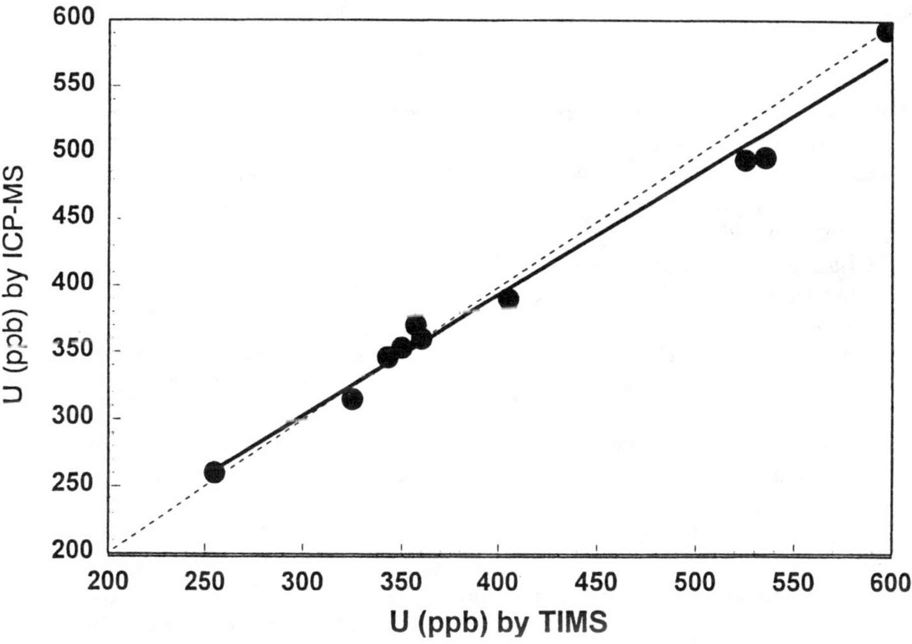

Figure 3 - Comparison of TIMS and ICPMS Results for Uranium Levels in Carbonate Rock Samples

For example, avid consumers of "mineral water" may have uranium intakes that vary by several orders of magnitude, according to the uranium content in their favorite brand. The uranium content in the drinking water in Israel is indicative of their source: local wells or Lake Kinneret. In wells and drillings, the uranium content (in combination with other indicators) may serve as a means of identifying the aquifer from which water is drawn. Finally, the uranium content in carbonate rocks was also determined by a relatively simple procedure, and the results obtained by ICPMS correlated well with isotope dilution-TIMS results.

References

[1] Denoyer, E. R.," An Advanced ICPMS Instrument," *International Laboratory,* April 1995, pp. 8-13.

[2] Lorber, A., Karpas, Z. and Halicz, L., "Flow Injection Method for Determining Uranium in Urine and Serum by ICPMS," *Analytica Chimica Acta*, Vol. 334, 1996, pp. 295-301.

[3] Twiss, P., Watling, R. J. and Delev, D. "Determination of Thorium and Uranium in Faecal Material from Occupationally-Exposed Workers Using ICP-MS," *Atom. Spectrosc.*, Vol. 14, 1994, pp. 36-39

[4] Ting, B. G., Paschal, D. C. and Caldwell, K. L., "Determination of Thorium and Uranium in Urine with Inductively Coupled Argon Plasma Mass Spectrometry," *Anal. Atom. Sci.* Vol. 11, 1996, pp. 339-342.

[5] Karpas, K., Halicz, L., Roiz, J., Marko, R., Katorza, E., Lorber, A. and Goldbart, Z., "Inductively Coupled Plasma Mass Spectrometry as a Simple, Rapid and Inexpensive Method for Determination of Uranium in Urine and Fresh Water: Comparison with LIF," *Health Physics*, Vol. 71, 1996, pp. 877-883.

[6] Lorber, A., Halicz, L., Karpas, Z., Elish, E., Roiz, J. and Marko, R., "Uranium in Urine and Serum of "Normal" Populations: a FIAS-ICPMS Study," in *"Plasma Source Mass Spectrometry: Developments and Applications"*, G. Holland and S. D. Tanner, Eds., The Royal Society. of Chemistry, United Kingdom, 1997.

[7] Karpas, Z., Lorber, A., Elish, E., Marcus, P., Roiz, J., Marko, R., Kol, R. Brikner, D. and Halicz, L., "Uranium in Urine - Normalization to Creatinine," *Health Physics,* Vol. 74, 1998, pp. 86-90.

[8] Karpas, Z., Lorber, A., Elish, E., Kol, R., Roiz, J., Marko, R., Katorza, E., Halicz, L., Riondato, J., Vanhaecke, F. and Moens, L., "The Uptake of Ingested Uranium After Low Acute Intake, " *Health Physics,* Vol. 74, 1998, pp. 337-345.

[9] Boeniger, M. F., Lowry, L. K., and Rosenberg, J., "Interpretation of Urine Results Used to Assess Chemical Exposure with Emphasis on Creatinine Adjustment: A Review," *American Industrial Hygiene Association Journal,* Vol. 54, 1994, pp. 615-627.

[10] Byrne, A. R. and Benedik, L, "Urinary Excretion of Uranium and Dietary Intake", *Health Physics,* Vol. 68, 1995, pp. 732-733.

[11] Halicz, L., Gavrieli, I, McLaren, J. W. and Lam, J. W. H., "Trace Metal Concentrations in the Dead Sea Brine - a Preliminary Study," *Geological Survey of Israel, Current Research,* Vol. 9, 1994, pp. 1-4.

[12] Halicz, L., Bar-Matthews, M., Ayalon, A. and Kaufman, A., "Determination of Low Concentrations of U and Th in Carbonate Rocks Using FI-ICPMS," *Atom. Spectrosc.* Vol. 18, 1997, pp. 175-179.

Linda A. Lewis[1] and George K. Schweitzer[2]

^{99}Tc BIOASSAY: A DIRECT COMPARISON OF LIQUID SCINTILLATION RADIATION DETECTION AND ICP-MS MASS DETECTION OF THE ^{99}Tc ISOTOPE

REFERENCE: Lewis, L. A. and Schweitzer, G. K., "**^{99}Tc Bioassay: A Direct Comparison of Liquid Scintillation Radiation Detection and ICP-MS Mass Detection of the ^{99}Tc Isotope,**" *Applications of Inductively Coupled Plasma-Mass Spectrometry to Radionuclide Determinations: Second Volume, ASTM STP 1344*, R.W. Morrow and J. S. Crain, Eds., American Society for Testing and Materials, 1998.

ABSTRACT: A means of analyzing ^{99}Tc in urine by inductively coupled plasma mass spectrometry (ICP-MS) has been developed. Historically, ^{99}Tc analysis was based on the radiometric detection of the 293 keV $E_{Max.}$ beta decay product by liquid scintillation or gas flow proportional counting. A separation scheme for ^{99}Tc detection by ICP-MS is given and is proven to be a sensitive and robust analytical alternative. A comparison of methods using radiometric and mass quantitation of ^{99}Tc has been conducted in water, artificial urine, and real urine matrices at activity levels between 700 and 2200 dpm/L. Liquid scintillation results based on an external standard manual quench correction and an automatic quench curve correction method are compared with results obtained by ICP-MS. Each method produced accurate results; however, the precision of the ICP-MS results is superior to that of liquid scintillation results.

KEYWORDS: ^{99}Tc, bioassay, urine, ICP-MS, liquid scintillation

---

To date, 21 isotopes of technetium with half-lives ranging from one second to several million years are known to exist. Technetium (atomic number 43) is the lightest of the artificial radioelements. ^{97}Tc ($t_{1/2}$ = 2.6E6 y), ^{98}Tc ($t_{1/2}$ = 4.2E6 y), and ^{99}Tc ($t_{1/2}$ = 2.13E5 y) are the only isotopes of technetium with half-lives over 90 days. Of the three longer lived isotopes, only ^{99}Tc has been obtained in weighable quantities because of the high (6.1%) fission yield [*1*]. ^{99}Tc is produced either

---

[1] Development Chemist, Analytical Services Organization, Lockheed Martin Energy Systems, Inc., 113C Union Valley Road, Oak Ridge, TN 37831-8244.

[2] Department of Chemistry, University of Tennessee, Knoxville, TN 37996-1600.

directly as a fission product of ^{235}U and ^{239}Pu or indirectly as a decay product of ^{99}Mo. The release of ^{99}Tc into the environment has occurred predominantly through reprocessing of nuclear fuel and testing of nuclear weapons, and to a lesser extent through use of radiopharmaceuticals and radioactive waste site contamination. Once released into the environment, it may exist as the volatile heptoxide (Tc_2O_7) or as the highly soluble and mobile pertechnetate ion (TcO_4^-). In these forms, the potential exists for technetium to be concentrated in plants and animals [*2*]. In humans, the pertechnetate ion localizes predominantly in the thyroid gland and gastrointestinal tract [*3*].

At the formerly named Oak Ridge Gaseous Diffusion Plant (ORGDP), ^{99}Tc was introduced into the gaseous diffusion cascades as a contaminant in uranium that had been reprocessed from spent nuclear fuel [*4*]. From a radiological protection perspective, it is imperative to control/monitor worker exposure during disassembly and decontamination processes using accurate evaluation techniques. To date few studies on ^{99}Tc bioassay have been reported. The majority of the ^{99}Tc literature reports the evaluation of ^{99}Tc in environmental, waste, and geochemical samples using both radioanalytical and mass analysis [*5-8*].

In a urine matrix, the analysis of ^{99}Tc is plagued with many difficulties using conventional radiometric methods. Difficulties originate during chemical separation due to the volatile nature of Tc_2O_7 or during radiation detection due to color or chemical quenching. In order to compare the ICP-MS detection capabilities to that of a radioanalytical technique, a method was developed that enabled samples to be analyzed by both detection methods. To accomplish this goal, a procedure was designed to remove both radioanalytical interferents (^{40}K and ^{99}Mo), as well as mass interferents (^{99}Mo, ^{99}Ru, and dissolved solids).

Experimental Method

Reagents

In each stage of the research project, 18 MΩ water was used. Nitric and hydrochloric acid solutions were prepared using commercially available high purity concentrated acids. A strongly basic anion exchange resin was used in the separation procedure and prepared by soaking in water for a minimum of 10 hours. Artificial urine was prepared according to the recipe described in Reference 9. All reagents used to prepare the artificial urine and the hydrogen peroxide were analytical grade. INSTA-GEL XF® (Packard Instrument Co., Meriden, CT) liquid scintillation cocktail was chosen for this project due to its ability to better handle the quenching characteristics of a prepared urine sample matrix. A 1000 μg/mL spectrometric standard solution of indium was obtained commercially. A ^{99}Tc standard was procured from the National Institute of Standards and Technology (NIST). Working solutions of the ^{99}Tc standard were prepared by directly weighing the materials, and secondary standard activities were verified by liquid scintillation counting.

Sample Preparation

After evaluating experimental data, a ^{99}Tc separation method was devised that eliminated interferents for both liquid scintillation and ICP-MS analysis. Composite water, artificial urine, and real urine samples were spiked at three different levels with ^{99}Tc. Ten 50 mL samples of each level/matrix were pipetted into 250 mL beakers. The samples were digested with 5 mL H_2O_2 in a covered beaker on a hotplate at a surface temperature between 150 to 175°C. After approximately one hour, the samples were removed from the hotplate and cooled. After cooling, the watch glass coverings were rinsed with 0.50 M HNO_3, and the samples were acidified with 2.5 mL of concentrated HNO_3 and diluted to 80 mL. The samples were warmed in order to aid in dissolving slightly soluble salts, cooled to room temperature, and poured over columns containing 1.8 mL of AG MP-1 anion exchange resin that had been prepared by rinsing with 30 mL of 0.50 M HNO_3. The sample flow rate was approximately 1 mL/min. Samples were then washed sequentially with 30.0 mL volumes of 0.50 M HNO_3 and 0.50 M HCl, eluted with 30 mL 3.0 M HNO_3 (heated to 95 °C), and then diluted to 50 mL. At this point, each sample was divided into two 25 mL portions. Samples to be analyzed by liquid scintillation were evaporated to approximately 1 mL volumes for acid removal, transferred to a 22 mL polyethylene liquid scintillation vial, and diluted to an 8 mL mark. Liquid scintillation sample preparation was completed by the addition of 12.0 mL of cocktail. Samples were well mixed and counted for a period of one hour. ICP-MS samples were transferred to 10 mL vials, spiked with a 0.2 ppb ^{115}In internal standard, and analyzed in peak-jumping mode. The remaining ICP-MS prepared sample solutions were stored for a peroid of one month for the purpose of reanalysis.

The use of a yield monitor was not employed in this study; however, direct percent bias and recovery were evaluated for each matrix and level.

Instrumentation

A VG Elemental PlasmaQuad PQ 2+, equipped with a De Galan V-grove nebulizer, was used for the ICP-MS analysis. Initially ICP-MS operating parameters were optimized for the analysis of ^{99}Tc. The conditions that were chosen are listed in Table 1. Each week, or after an instrumental modification was made, a mass calibration was conducted. Prior to each analysis, a tuning standard containing 10 ppb each of Be, Mg, Co, Ni, In, Ce, Pb, and U was used to adjust sensitivity, focus the plasma, check mass calibration, and maximize resolution. Once instrumental operating conditions were established, a short term stability check was conducted. The stability check was carried out by observing 10 consecutive 30 s analyses of the tune solution. From mass countrates, the means, standard deviations, and relative standard deviations were determined. Less than 2% relative standard deviation was the goal for the higher mass elements In, Pb, and U.

Liquid scintillation samples were corrected for quench using both the manual external standard and automatic quench curve corrections. Liquid Scintillation analysis was conducted on a Packard 2500TR analyzer with low level counting capabilities. In order to prepare liquid scintillation counters for automatic quench

TABLE 1--*Instrumental Operating Parameters.*

ICP-MS	
Forward Power/W	1350
Coolant Gas Flow Rate/L min^{-1}	14.5
Auxiliary Gas Flow Rate/ L min^{-1}	0.94
Nebulizer Gas Flow Rate/ L min^{-1}	0.90
Sample Solution Flow-rate/ mL min^{-1}	0.92
Interface	
Sample Cone	Nickel
Skimmer Cone	Nickel
Acquisition Parameters	
Acquisition Time/s	45
Dwell Time/ms	10.24
Time per Sweep/s	0.60
Points per Peak	19
DAC per Point	9
Scan Mode	Peak-jumping
Signal Processing	Spectral Peaks Integrated
Detection Mode	Pulse Counting
Internal Standard	^{115}In
Vacuum Pressures	
Expansion/mbar	1.9×10^{0}
Interface/mbar	$<1.0 \times 10^{-4}$
Analyzer/mbar	1.9×10^{-6}

correction, a quench curve was established by spiking a series of ^{99}Tc standards at a constant activity in the appropriate liquid scintillation cocktail with increasing amounts of a quenching agent such as acetone.

^{3}H, ^{14}C, and background Instrument Performance Standards (IPA) were counted before each sample batch to monitor counter efficiency and background. Samples prepared for liquid scintillation counting were placed in the counters, dark adapted for five minutes, and counted sequentially for one hour.

Calculations

Two detection limits were calculated for both detection methods. A method limit of detection (LOD) was calculated from the ICP-MS data using the standard deviation (s_B) of ten method blanks analyzed over a period of one week:

$$LOD = 3.29(s_B) \quad (1)$$

The minimum detectable concentration (MDC) was the second detection limit determined. A sample activity above the MDC represents a positive radiation dose within a 95% confidence level. MDC is based upon the lowest possible activity per unit volume of an analyte in urine which may be detected by a particular preparation and detection method. An MDC s_B represents a propagated value that incorporates the analytical blank and an unexposed person's activity concentration. The unexposed person's blank activity standard deviation was assumed equivalent to the analytical blank due to a lack of a real urine blank population for ^{99}Tc. The ICP-MS MDC was calculated using the following equation:

$$MDC = 4.65\ s_B \quad (2)$$

The activity per unit volume for samples counted by liquid scintillation was calculated using the following equation:

$$dpm/V = (S_{cnt} - B_{cnt})/\ K \quad (3)$$

where S_{cnt} and B_{cnt} were the counts in the 2 to 294 keV regions for the sample and blank, and K represents e (efficiency), T (time), and V (volume). Efficiency was determined both automatically using a quench curve and manually using the following equation:

$$e = Q_s/s \quad (4)$$

Q_s was calculated by subtracting a samples quenched counts including background ($Q_U + Q_B$) from that sample's quenched counts and background after being exposed to a ^{133}Ba source ($Q_s + Q_U + Q_B$). The value s was determined by subtracting an unquenched blank (prepared using deionized water and cocktail) and background (b) from that unquenched blank and background after being exposed to the ^{133}Ba source (s + b).

The LOD for liquid scintillation counting used a similar equation to the ICP-MS LOD equation:

$$LOD = 3.29(s_B)/K \quad (5)$$

where the blank standard deviation (s_B) is expressed as counts. A modified equation is used to calculate MDC:

$$MDC = 4.65\ s_B/K + 3/K \quad (6)$$

This MDC equation is stated in the ANSI N13.30 Standard for radiobioassay [10].

Results

Water, artificial urine, and real urine matrices spiked at three different levels, approximately 700, 1420, and 2120 dpm/L, were evaluated. Samples were labeled according to matrix and activity level, 1 lowest and 3 highest activity, using W1, W2, W3, A1, A2, A3, R1, R2, and R3 identifiers. Ten samples of each matrix and level were taken through the ^{99}Tc separation procedure as a group, and the final 50 mL solutions were divided into two aliquots, one for each analysis method. Sample results were not corrected for yield; however, the percent ^{99}Tc recovery and bias were evaluated for each matrix and level.

Tables 2 - 4 contain the liquid scintillation (manual and automatic quench corrected) and ICP-MS results. Evaluation of the compiled data revealed that both liquid scintillation manual and automatic quench corrected standard deviations tended to be higher than by ICP-MS. The major difference in s_B values was noted in the low level W1, A1, and R1 sets in which the liquid scintillation deviations were 1.4 to 6.5 times higher than that of the ICP-MS values. The urine matrix s_b revealed that the largest extreme between radiation counting (83.8 and 87.7 dpm/L) and mass detection (13.0 dpm/L) existed at the 700 dpm/L level. The liquid scintillation counting error was too large to make a direct sample per sample comparison with corresponding ICP-MS data.

%Bias was calculated and included in Tables 2 - 4. For each matrix and level, the manual quench corrected liquid scintillation data was closer to the true value than that of the automatic quench corrected liquid scintillation and ICP-MS data. Experimental liquid scintillation data for the manual and automatic quench corrected results revealed that the manual method more effectively corrected for color quenching.

Reagent blanks were evaluated for each matrix and detection method (Tables 5-7) in order to determine the minimum detectable concentration (MDC) and limit of detection (LOD) values. The manual quench efficiencies were used to detemine the liquid scintillation detection limits since the values accurately reflected both chemical and color quenching. The larger standard deviation in total blank counts obtained by liquid scintillation had a detrimental effect on the calculated detection limits. Liquid scintillation MDC and LOD values were 2, 5, and 13 times higher for water, artificial urine, and real urine matrices, respectively, than ICP-MS values. The real urine MDC was 263.3 dpm/L for liquid scintillation counting and 20.74 dpm/L for ICP-MS, and the LOD was 203.4 dpm/L for liquid scintillation counting and 14.67 dpm/L for ICP-MS.

TABLE 2--^{99}*Tc Water Comparison*

Control ID	Liquid Scintillation Manual Q (dpm/L)	Liquid Scintillation Q Curve (dpm/L)	ICP-MS (dpm/L)
W1-1	725.4	672.4	583.4
W1-2	602.7	540.4	625.6
W1-3	654.7	592.4	589.5
W1-4	632.4	626.0	652.6
W1-5	733.4	606.8	652.9
W1-6	711.3	639.6	649.3
W1-7	619.0	550.0	624.9
W1-8	731.1	664.4	652.0
W1-9	657.7	576.4	652.3
W1-10	632.9	588.0	638.1
Mean	670.1	605.6	632.1
s_b	50.4	44.9	26.4
%Bias	-5.9	-15	-11
W2-1	1226	1200	1196
W2-2	1273	1201	1259
W2-3	1329	1276	1273
W2-4	1246	1200	1266
W2-5	1247	1178	1231
W2-6	1224	1178	1260
W2-7	1161	1096	1193
W2-8	1164	1122	1189
W2-9	1250	1208	1214
W2-10	1233	1174	1268
Mean	1235	1183	1235
s_b	48.8	49.0	34.4
%Bias	-13	17	-13
W3-1	1715	1956	1866
W3-2	2066	2087	1989
W3-3	1962	1979	1964
W3-4	1885	1901	1900
W3-5	2055	2069	1837
W3-6	1979	1999	2018
W3-7	1719	1747	1635
W3-8	1813	1873	1926
W3-9	2139	2182	1974
W3-10	2112	2100	1913
Mean	1945	1989	1902
s_b	156	128	109
%Bias	-8.9	-6.8	-11

TABLE 3--^{99}*Tc Artificial Urine Comparison*

Control ID	Liquid Scintillation Manual Q (dpm/L)	Liquid Scintillation Q Curve (dpm/L)	ICP-MS (dpm/L)
A1-1	532.6	417.9	562.9
A1-2	550.5	477.9	602.1
A1-3	481.8	392.7	533.3
A1-4	561.5	462.7	631.5
A1-5	538.6	358.7	558.8
A1-6	495.2	448.3	591.3
A1-7	974.2	513.1	604.4
A1-8	613.2	520.7	613.5
A1-9	-	-	633.0
A1-10	623.1	534.3	559.6
Mean	596.7	458.5	589.0
s_b	149	60.2	33.8
%Bias	-16	-36	-17
A2-1	1540	1170	1258
A2-2	1065	918	1323
A2-3	1436	1216	1296
A2-4	1565	1237	1305
A2-5	1344	1186	1328
A2-6	1340	1192	1306
A2-7	1128	1082	1307
A2-8	1181	1142	1379
A2-9	1149	1274	1353
A2-10	1356	1174	1219
Mean	1310	1159	1307
s_b	174	99.4	45.2
%Bias	-7.9	-18	-8.1
A3-1	2104	1967	1903
A3-2	2036	1896	1836
A3-3	2125	1964	1969
A3-4	2063	1920	1913
A3-5	2071	1980	1932
A3-6	2003	1918	1951
A3-7	2075	1846	1936
A3-8	1962	1758	2033
A3-9	1886	1844	1909
A3-10	2082	1913	1950
Mean	2041	1901	1933
s_b	72.3	68.6	50.8
%Bias	-4.4	-11	-9.4

TABLE 4--^{99}*Tc Real Urine Comparison*

Control ID	Liquid Scintillation Manual Q (dpm/L)	Liquid Scintillation Q Curve (dpm/L)	ICP-MS (dpm/L)
R1-1	627.9	526.2	630.3
R1-2	739.0	614.2	612.7
R1-3	593.4	500.6	587.0
R1-4	507.0	491.0	614.1
R1-5	581.7	602.6	604.8
R1-6	489.9	458.2	598.6
R1-7	580.2	561.4	611.3
R1-8	717.9	685.4	617.5
R1-9	641.3	598.6	626.4
R1-10	693.9	734.6	621.5
Mean	617.2	577.3	612.4
s_b	83.8	87.7	13.0
%Bias	-12	-18	-13
R2-1	1764	1308	1239
R2-2	1433	1232	1200
R2-3	1535	1367	1223
R2-4	1552	1268	1273
R2-5	1230	1203	1266
R2-6	1539	1336	1329
R2-7	1576	1339	1362
R2-8	1532	1305	1316
R2-9	1293	1227	1316
R2-10	1415	1224	1309
Mean	1487	1281	1283
s_b	152	57.6	51.8
%Bias	4.8	-9.7	-9.6
R3-1	2239	1948	1985
R3-2	1899	1733	1802
R3-3	2065	1814	1982
R3-4	2040	1904	1861
R3-5	2021	1834	1910
R3-6	2241	2012	1977
R3-7	2236	1897	2083
R3-8	2211	1738	1936
R3-9	2441	1952	1971
R3-10	2390	2026	2045
Mean	2178	1886	1955
s_b	170	104	82.5
%Bias	2.7	-11	-7.9

TABLE 5-- ^{99}*TC Detection Limits In Water*

Sample ID	ICP-MS (dpm/L)	Liquid Scintillation (Counts)
WB1	180.7	1628
WB2	151.6	1601
WB3	158.0	1581
WB4	152.6	1579
WB5	143.3	1654
WB6	214.3	1513
WB7	172.5	1558
WB8	177.8	1644
WB9	179.0	1653
WB10	177.3	1677
WB11	189.7	1580
WB12	169.3	1513
WB13	154.0	1584
WB14	157.0	1602
WB15	168.0	1557
s_b	18.10	49.43
MDC (dpm/L)	84.17	190.9
LOD (dpm/L)	59.54	162.6

Liquid Scintillation Parameters: e = 0.81, Volume = 0.025 mL, Time = 60 min

TABLE 6-- ^{99}*TC Detection Limits In Artificial Urine*

Sample ID	ICP-MS (dpm/L)	Liquid Scintillation (Counts)
AB1	-	1628
AB2	112.1	1601
AB3	93.93	1581
AB4	84.19	1579
AB5	93.53	1654
AB6	106.1	1513
AB7	96.76	1558
AB8	87.86	1644
AB9	104.1	1653
AB10	90.25	1677
AB11	107.0	1611
AB12	94.47	1599
AB13	99.25	1627
AB14	93.88	1642
AB15	84.92	1682
s_b	8.49	47.70
MDC (dpm/L)	39.57	203.8
LOD (dpm/L)	28.00	156.9

Liquid Scintillation Parameters: e = 0.74, Volume = 0.025 mL, Time = 60 min

TABLE 7-- *^{99}TC Detection Limits In Real Urine*

Sample ID	ICP-MS (dpm/L)	Liquid Scintillation (Counts)
RB1	70.84	1596
RB2	68.07	1624
RB3	74.97	1644
RB4	66.20	1620
RB5	68.23	1593
RB6	70.38	1461
RB7	78.31	1541
RB8	70.12	1592
RB9	68.65	1534
RB10	79.15	1675
s_b	4.46	61.81
MDC (dpm/L)	20.74	263.3
LOD (dpm/L)	14.67	203.4

Liquid Scintillation Parameters: e= 0.74, Volume = 0.025 mL, Time = 60 min.

The overall average percent of the known value for each matrix and activity level was consistently higher for the manual correction method. The average percent of known by liquid scintillation manual quench, liquid scintillation quench curve, and ICP-MS were: 90%, 87%, and 88% in water; 90%, 77%, and 88% in artificial urine; and 98%, 87%, and 89% in real urine.

Conclusion

Based on the population of ten values per matrix and level, the manual quench corrected liquid scintillation overall %Bias was lower than the automatic quench corrected and ICP-MS values for each matrix and level. However, on a per-sample basis, the error on the manual quench corrected result was much greater than that of ICP-MS. For the ^{99}Tc isotope, the larger per-sample error for radiation detection compared with mass detection stems from the decreased counting statistics when observing decay products rather than atoms. This is generally true for the long-lived isotopes (greater than 1000 years); however, radiation detection is still a more sensitive method of analysis for the shorter-lived isotopes. In a real urine matrix, limits of detection (LOD) for ICP-MS and liquid scintillation detection are 14.67 and 203.4 dpm/L, respectively. Because of the better precision for ^{99}Tc by ICP-MS, lower limits of detection and similar accuracy's were achievable compared to the radiation detection method. Other advantages of ^{99}Tc analysis by ICP-MS compared to liquid scintillation detection include the elimination of a mixed hazardous waste, and a 50% reduction in turnaround time resulting in a 40% reduction in analysis cost.

Acknowledgment

This work was done under DOE contract DE-AC05-84OR21400 for the U. S. Department of Energy.

References

[*1*] Crouch, E. A. C., *Atomic Data Nuclide Data Tables*, Vol. 19, p. 440, 1977.

[*2*] Hurtgen, C., Koch, G., van der Ben, D., Bonotto, S., "The Determination of Technetium-99 in the Brown Marine Alga Fucus Spiralis Collected Along the Belgian Coast," *Science of the Total Environment*, Vol. 70, pp. 131-142, 1988.

[*3*] J. E. Till, F. O. Hoffman, D. E. Dunning, Jr., "A New Look at ^{99}Tc Releases to the Atmosphere," *Health Physics*, Vol. 36, pp. 21-30, 1979.

[*4*] Simmons, D. W., "An Introduction to Technetium in the Gaseous Diffusion Cascades," Report KTSO39, Government Printing Office, Washington, DC, 1996.

[*5*] Anderson, T. J., Walker, R. L., " Determination of Picogram Amounts of Technetium-99 by Resin Bead Mass Spectrometric Isotope Dilution," *Analytical Chemistry*, Vol. 52, pp. 709-713, 1980.

[*6*] Verrezen, F., Hurtgen, C., "The Measurement of Technetium-99 and Iodine-129 in Waste Water from Pressurized Nuclear-power Reactors," *Applied Radiation and Isotopes*, Vol. 43, pp. 61-68, 1992.

[*7*] Erickson, M. D., Aldstadt, J. H., Alvarado, J. S., Crain, J. S., Orlandini, K. A., Smith, L. L., "Radiochemical Method Development," *Journal of Hazardous Materials,* Vol. 41, pp. 351-358, 1995.

[*8*] Beals, M. D., "Determination of Technetium-99 in Aqueous Samples by Isotope Dilution Inductively Coupled Plasma - Mass Spectrometry," *Journal of Radioanalytical and Nuclear Chemistry,* Vol. 204, 1996, pp. 253-263.

[*9*] Robinson, A. V., Fisher, D. R., Hadley, R. T., "Technical Evaluation of Draft ANSI Standard N13.30, Performance Criteria for Radioassay," Report No. PNL-5107, Pacific Northwest Laboratory, Richland, WA, 1984.

[*10*] *American National Standard Performance Criteria for Radiobioassay*, Health Physics Society, ANSI HPS N13.30, American National Standards Institute, Inc., New York, 1996.

Christopher J. Pickford,[1] Janice Haines,[1] Ruth Hearn[1] and John McAughey[2]

DETERMINATION OF RADIONUCLIDES IN BIOLOGICAL FLUIDS USING A HIGH RESOLUTION ICP-MS IN LOW RESOLUTION MODE.

REFERENCE: Pickford, C.J., Haines, J., Hearn, R. and McAughey, J. **"Determination of Radionuclides in Biological Fluids Using a High Resolution ICP-MS in Low Resolution Mode,"** *Applications of Inductively Coupled Plasma-Mass Spectrometry to Radionuclide Determinations: Second Volume, ASTM STP 1344*, R.W. Morrow and J.S. Crain, Eds., American Society for Testing and Materials, 1998.

ABSTRACT: High Resolution (HR) ICP-MS in low resolution mode was used with ultrasonic nebulization to determine long-lived radionuclides and surrogates in human body fluid samples. Although very sensitive, achieving LODs less than 0.1 fg/mL for ^{244}Pu, a number of molecular ion interferences have been encountered from commonly-occurring species which limit sensitivity and require the development of specialized sample preparation routes.

KEYWORDS: ICP-MS, high resolution, trace analysis, biological samples, radiochemical methods, plutonium, lanthanoids

---

ICP-MS based techniques have been widely used for detecting radionuclides in environmental and other sample types, particularly where the half life is sufficiently long [1-3]. To reach levels of detection that are competitive with traditional nuclear spectrometric techniques, extensive sample preparation is often a prerequisite, both to concentrate the species to measurable levels (pg/mL) and to remove interfering species - mainly the high salt content likely in most environmental or biofluid samples [4]. In previous work, we have used Quadrupole ICP-MS in this way, both as a detector for environmental radionuclides [5] and as a means of measuring concentrations of radionuclides and

---

[1]R & D manager, manager - Inorganic Analysis, and senior analyst, respectively, AEA Technology/ Analytical Services, B551 Harwell, Didcot, Oxfordshire OX11 0RA England.

[2] Manager, Health-Related Aerosols, AEA Technology, B551 Harwell, Didcot, Oxfordshire OX11 0RA England.

surrogates used in human uptake studies [6-8]. In this case, ethical considerations severely limit the amounts of these species which can be administered to volunteers, and highly sensitive methods of analysis are therefore essential.

Carrying out such studies with actinide surrogates such as lanthanoids (or rare earths) is attractive in respect of their low toxicity, but may be of limited applicability because of the levels of chemical blank likely to be encountered if large volumes of, for example, urine are preconcentrated and separated to 1 - 2 mL of a solution ready for analysis; in a previous study, blank contamination effectively placed a limitation upon the lowest levels of certain lanthanoid surrogates which could be seen in human body fluids, and attempts to reach the "natural" background levels of these species for comparison purposes obtained levels which were probably unrealistically high in some cases [7]. Other published data for background levels of lanthanoids in human body fluids, obtained using different analytical techniques, may also be of limited value for the same reason [9].

High resolution ICP-MS is a relatively new technique which uses a double focusing mass spectrometer to separate the atomic ions formed in an ICP discharge [10], rather than a quadrupole MS. In comparison with Quadrupole ICP-MS instruments, HR ICP-MS is capable of much greater sensitivity if it is used in low resolution mode [8]; alternatively, HR ICP-MS can be used in high resolution mode, achieving resolution up to 10 000, compared with the 400 or so achieved by quadrupole instruments [11].

Low detection limits (fg/mL) are clearly most attractive for elements for which the natural background levels are likely to be very low; these tend to be groups of elements such as the lanthanoids, the platinum group metals and the actinides.

In the present work, a HR ICP-MS spectrometer in low resolution mode has been combined with an ultrasonic nebulizer and sensitive preconcentration and cleanup techniques to develop analytical methods for a number of analogues of $^{239/240}$Pu to be determined in the blood and urine of human volunteers participating in uptake studies. Analogues reported in this study comprise ^{244}Pu, (which is used as a relatively low radiotoxicity isotope of Pu) and the lanthanoid group of elements which have been used in previous studies as actinide analogues.

Experimental Method

High Resolution Inductively Coupled Plasma-Mass Spectrometer (ICP-MS)

The high resolution ICP-MS used was a PlasmaTrace 2, manufactured by Micromass (UK) Ltd. This was fitted with a CETAC USN 6000 ultrasonic nebulizer, operated at a solution uptake rate of 2.0 mL/min. The instrument was operated in a Class 10 000 cleanroom; parts of the sample preparation procedure were carried out in an adjacent Class 10 cleanroom.

The measurement conditions in Table 1 were used to quantify ^{244}Pu and lanthanoid concentrations. All measurements were carried out at a resolution of 400.

TABLE 1--*Scan and measurement parameters for the PlasmaTrace 2.*

Species Monitored	Dwell Time (ms)	Points/Peak	Scans
^{242}Pu	20	30	1
^{244}Pu	30	30	10
Lanthanoids (^{140}Ce to ^{175}Lu)	25	26	1

Standardization

^{244}Pu was quantified by isotope dilution with a known spike of ^{242}Pu, which was added to samples before processing. Aqueous calibration data were obtained without any sample preparation, in order to confirm the linearity of response. Concentrations were calculated by area ratioing.

Lanthanoids were quantified by calibration against aqueous standards after normalization of the analyte response using indium as an internal standard. No correction was made for recovery as this was considered to be close to 100 % (see below).

Sample Preparation

Urine samples (24 hour) for lanthanide analysis were stored at 5°C until ready for analysis (1-2 months). Each sample was carefully shaken to suspend deposited matter as thoroughly as possible and a 25 mL aliquot removed and transferred to a "Sterilin" 30 mL capped urine sample container. 500 pg of a Lu standard were added to each vial, followed by 2.0 mL of 1:1 H_2O/NH_4OH (Aristar). After standing overnight, the deposited phosphates (mainly Ca) were centrifuged into the conical base of the container, washed twice with ammonia solution and stored until ready for analysis. This procedure was essential in order to lower the total dissolved solids (TDS) levels of the urine samples, by removing sodium and potassium salts and soluble organic compounds.

Samples were then dissolved in 0.1 mL of high purity nitric acid (electronic grade), 25 ng of an indium standard added, and each solution diluted to 25 mL with high purity water immediately prior to analysis by HR ICP-MS. Blood samples were treated in the same way as urines, after an initial wet oxidation and ignition to 450°C.

Samples for ^{244}Pu determination were collected and stored in a cold room (5°C) without acid addition. The method of preparation used was closely based on a bioassay procedure used for the determination of Pu by Alpha Spectrometry [12]. Total 24 hour samples were spiked with ^{242}Pu yield tracer and digested with concentrated nitric acid. After addition of calcium and phosphate ions, solutions were made alkaline to precipitate calcium phosphate with Pu. The precipitated material was ignited at 450°C, hydrolyzed in dilute nitric acid to convert pyrophosphate to phosphate and adjusted to 8M with HCl. Pu was then absorbed onto a column of AG1X8 resin and then eluted with HI/HCl. After heating with nitric acid, the residue was made up to 2 mL with 0.2M nitric acid. After the interference from Pb had been identified, the procedure was altered to minimise the Pb content of the final measurement solution. High purity acids were used for digestion and the prepared 2 mL of solution passed down a column of Sr.Spec resin (Eichrom Industries Inc). This resin separates tetravalent Pu and divalent Pb nearly quantitatively in dilute nitric acid media, with the Pb being retained on the column [13]. Blood samples were treated in the same way as urine.

Results

Sensitivity and Limits of Detection (LODs)

Method limits of detection for blood and urine for ^{244}Pu and the lanthanoids are given in Table 2. The limits were calculated as three standard deviations (expressed as concentrations) of the method blank solutions which were run before and after each batch of analyzed samples; twelve blanks were measured for ^{244}Pu and a total of thirty two for the lanthanoids. Note that limits of detection refer to samples which have not been preconcentrated. The limit of detection of 0.07 fg/mL for ^{244}Pu refers to the 2 mL of analyte solution required for analysis. This translates to a theoretical limit of detection in a 24 hour sample of 0.14 fg, or approximately 0.1 attogram/mL in the original sample. For lanthanoids, the current sample preparation route for urine and blood results in a concentration factor of 1, i.e., detection limits in blood or urine are those in Table 2.

The detection limit for ^{244}Pu in urine (and blood, to a lesser extent) was initially worsened by the presence of interfering species (see below). Improving the sample preparation route and carrying out the latter stages of sample preparation in a Class 10 cleanroom removed these interferences.

Detection limits for the lanthanoids are effectively limited by the measurement procedure used and by the instrumental background - the "rare earths" (lanthanoids) are by no means rare at the pg level. Blank levels for urine for the more common lanthanoids are within a factor of two of the detection limit.

Calibration for all species was effectively linear over the measurement range of interest (e.g., Fig. 1) and up to the highest measurable concentrations, although if unidentified molecular ion interferences were present, apparent curvature at low concentration levels was seen.

TABLE 2--*Limits of detection.*

Analyte	Limit of Detection (fg/mL)
^{244}Pu	0.07
^{139}La	600
^{140}Ce	800
^{141}Pr, ^{146}Nd, ^{157}Gd, ^{163}Dy	100
^{149}Sm, ^{238}U	200
^{151}Eu, ^{175}Lu	20
^{159}Tb	30
^{165}Ho	10
^{166}Er	40
^{169}Tm	80
^{173}Yb	50

Measurement Precision

Measurement precision (RSD%) for replicate measurements for ^{244}Pu solutions in dilute acid were 3.7 % (100 fg/mL), 6.7 % (10 fg/mL) and 15 % (1 fg/mL). Precision for blood samples (10 mL) containing 1 fg/mL was 3.8 %, using the optimized procedure.

RSD% measured for the lanthanoids for an acid matrix surrogate solution analyzed together with the processed samples as an analytical QC material was 11 % (100 pg/mL) and 5 % (200 pg/mL). Comparable figures for solutions containing calcium and phosphate at the same levels as the processed urine samples were 14 % and 10 %.

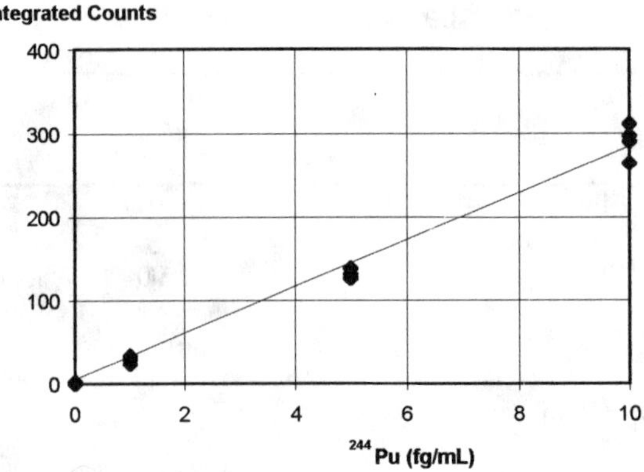

FIG. 1--*Calibration curve for* ^{244}*Pu in dilute acid matrix.*

RSD% measured for 20 pg/mL Lu added to a series of natural urines (29) carried through the sample preparation route was 9.1 %. These samples were analyzed to establish background levels of the lanthanoids for a normal human population.

Accuracy: Recovery of Added Analyte

^{244}*Pu:* For a level of 400 fg added to a 24 hour urine sample, a recovery of 102 % was obtained (after normalization using ^{242}Pu).

Lanthanoids: Mean recovery for the lanthanoids for the internal QC solutions (synthetic) which were analyzed together with a series of 120 natural urines gave values between 87 and 101 % (Table 3).

TABLE 3--*Recovery of internal QC solutions for lanthanoids*

QC Standard	Measured (pg/mL)	Recovery (%)
100 pg/mL Matrix Matched	101 ± 22	101
200 pg/mL Matrix Matched	194 ± 20	97
100 pg/mL Acid Standard	87 ± 18	87
200 pg/mL Acid Standard	182 ± 20	91

Mean recovery of the 20 pg/mL Lu spike for the series of urine samples was 22.9 (SD = 2.1) pg/mL (equivalent to 115 ± 10.5 %).

An estimate of the recovery of calcium in the urine preparation route for lanthanoids was obtained by measuring the concentration of Ca in the urine before and after clean-up, using ICP-AES. This gave an average recovery of 97.1 ± 2.1 %. As this was not significantly different from 100 % ($t=0.15$, $t_{cr}=1.96$, $P=0.05$), lanthanoid concentrations were not corrected for this factor.

Interferences

^{244}Pu: In initial work, it became apparent that there were substantial "blanks" for processed urine and blood samples. As no significant contamination of the reagents or laboratory environment with ^{244}Pu was likely, molecular ion interference was suspected.

The species initially suspected of causing the interference was $^{232}Th^{12}C$; significant levels of Th were found in processed urine solutions, as were carbon-containing species derived from the columns and eluents used.

Tests (Fig. 2) showed that even modest levels of both Th and C could give rise to apparent ^{244}Pu concentrations considerably more than real levels which would be encountered in the human uptake studies to be undertaken.

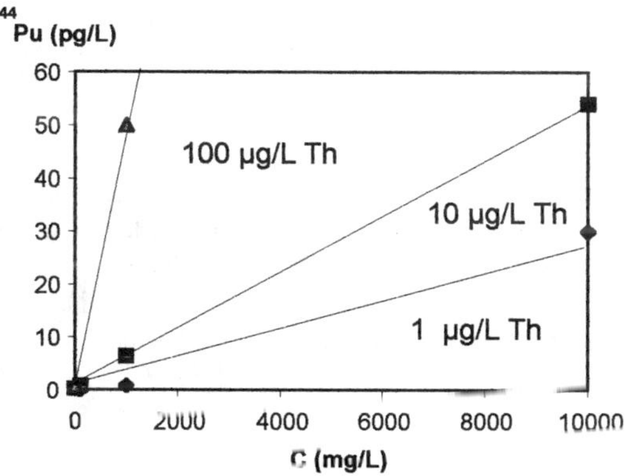

FIG. 2 *Apparent ^{244}Pu concentrations caused by $^{232}Th^{12}C$.*

The sample preparation route was extended to ensure that Th and C were eliminated as far as possible from the processed solutions, but some apparent interference still remained. Other species which might be expected to produce an ion at m/z 244 include $^{207}Pb^{37}Cl$ and $^{204}Pb^{40}Ar$. An examination of the spectrum of a typical prepared blank for urine confirmed that relatively large amounts of Pb were present, and that lead + argon and lead + chlorine species could be seen in some solutions, giving an apparent ^{244}Pu peak equal to a solution concentration of up to 35 fg/mL.

Carrying out part of the sample preparation in a Class 10 cleanroom, using electronic grade nitric acid and including a final digestion to remove chloride lowered blanks effectively to zero, and a limit of detection of 0.07 fg/mL was obtained.

Lanthanoids: At the levels measured, occasional elevated levels were seen for Er and Tb in urine; these did not appear to be real, as it is extremely unlikely that individual lanthanoids would appear in urine samples in this way, but to derive from an unidentified interference. Isotopes of other lanthanoids, not used for quantitation, also suffered from other unidentified interferences, as well as from BaO/BaOH interferences.

Conclusion

HR ICP-MS is a powerful tool for low level measurement of radionuclides and other elements for which chemical blanks are likely to be very low. However, there is a risk that molecular ion interferences will occur at certain mass numbers which are not necessarily easy either to predict or to eliminate. Sample preparation procedures need to be chemically clean, using pure reagents and a clean room environment if necessary.

In this work, a method has been developed for measuring ^{244}Pu levels in biological fluid samples at levels sufficiently low for the technique to be useful in human uptake studies, provided that rigorous efforts are made to control chemical blanks caused by interfering elements such as Th and Pb. The methodology developed for the lanthanoids is sensitive enough to reach natural background levels of these elements in blood and urine, enabling stable isotope spike intake and uptake to be estimated.

References

[1] Chiappini, R., Taillade, J.M. and Brebion, S., *J. Analytical Atomic Spectrometry*, Vol. 11, 1996, pp. 497-503.

[2] Crain, J.S., Smith, L.L., Yaeger, J.S. and Alvarado, J.A., *J. Radioanalytical Nuclear Chemistry*, Vol. 194, 1995, pp. 133-139.

[3] Pickford, C.J., Brown, R. and Long, S.E., "Measurement of Long-Lived Radionuclides by Non-Radiometric Methods," *Science of the Total Environment*, Vol. 70, 1988, pp. 265-274.

[4] Kershaw, P. J., Sampson, K.E., McCarthy, W and Scott, R.D., *J. Radioanalytical Nuclear Chemistry*, Vol. 198, 1995, pp. 113-124.

[5] Cox, R., Pickford, C.J. and Thompson, M., *J. Analytical Atomic Spectrometry*, Vol. 7, 1992, pp. 635-640.

[6] Dalgarno, B.G., Brown, R.M. and Pickford, C.J., *Biomedical & Environmental Mass Spectrometry*, Vol 16, 1988, pp. 377-380.

[7] McAughey, J., Vernon, L., Haines, J., Sanders, T., and Clark, R., *Proceedings of the 9th International Conference on Heavy Metals in the Environment*, Toronto, 1993, pp. 117-120.

[8] Cox, R. J., Pickford, C.J. and Hearn, R., "Measurement of Actinides Using High Resolution ICP-MS," *Winter Conference*, Cambridge, UK, 1996.

[9] Minoia, C., Sabbioni, E., Apostoli, P., Petra, R., Pozzoli, L., Gallorini, M., Nicolaou, G., Alessio, L. and Capodoglio, E., *Science of the Total Environment*, Vol. 95, 1990, pp. 89-105.

[10] Kim, C. K., Seki, R., Morita, S., Yamasaki, S. I., Tsumura, A., Takaku, Y., Igarashi,Y and Yamamoto, M., *J. Analytical Atomic Spectrometry*, Vol. 6, 1991, pp. 205-209.

[11] Cox, R.J., Pickford, C. J., Reed, N. M. and Thompson, M., *Proceedings 5th Surrey Conference on PSMS*, Durham (UK), 1993.

[12] Eakins, J.D., Lally, A.E., Morgan, A. and Sandalls, F.G., UKAEA Report AERE-AM 103, 1968.

[13] Analytical Products Description, Eichrom Industries Inc., Darien, Illinois 60561, U.S.A., 1995.

Michael E. Ketterer[1] and Christopher J. Khourey[2]

HIGH-PRECISION DETERMINATION OF $^{234}U/^{238}U$ ACTIVITY RATIOS IN NATURAL WATERS AND CARBONATES BY ICPMS

REFERENCE: Ketterer, M.E. and Khourey, C.J., **"High-Precision Determination of $^{234}U/^{238}U$ Activity Ratios in Natural Waters and Carbonates by ICPMS,"** *Applications of Inductively Coupled Plasma-Mass Spectrometry to Radionuclide Determinations: Second Volume, ASTM STP 1344*, R.W. Morrow and J.S. Crain, Eds., American Society for Testing and Materials, 1998.

ABSTRACT: A method has been developed for precise measurement of $^{234}U/^{238}U$ activity ratios in natural waters and carbonates using quadrupole inductively coupled plasma mass spectrometry. A recovery of 80-85% of seawater U is achieved by Fe(III) coprecipitation followed by extraction chromatography with a supported dipentyl pentane phosphonate material; 90-95% of U is recovered from carbonates, which are are dissolved in HNO_3 and subjected to the same extraction chromatographic preparation. Isotopic measurements are made via recirculating pneumatic nebulization of small volumes of solutions containing 0.5-5 mg/L U. $^{234}U/^{235}U$ is measured as a proxy for determination of $^{234}U/^{238}U$; iridium is added to sample solutions and the ion ratio $^{191}Ir^{40}Ar^+/^{193}Ir^{40}Ar^+$ is measured for internal mass discrimination correction. $^{234}U/^{238}U$ activity ratios in the range 1.143-1.154 are observed for 13 seawater and contemporary corals, in agreement with the established marine $^{234}U/^{238}U$ activity ratio. For sample sizes of 5-25 μg U, ICPMS uncertainties of +/- 0.2-0.5 % relative, 2σ standard error, approach those obtained for < 0.1 μg U by thermal ionization mass spectrometry. Measurements of $^{234}U/^{238}U$ activity ratios in bottled waters, Lake Erie surface waters, mollusk fossils, and fertilizers are also demonstrated.

KEYWORDS: environmental, water, carbonates, ICPMS, isotope ratio MS, uranium, extraction chromatography

---

The differences in physical and chemical properties of a nuclide such as ^{238}U and its radiogenic daughters may bring about mutual separations of the parent and daughters; this gives rise to a "disequilibrium" condition. Where disequilibrium exists,

(1) Department of Chemistry, John Carroll University, University Heights, OH 44118.
(2) Department of Geological Sciences, Case Western Reserve University, Cleveland, OH 44106.

the following relationship is invalid:

$$N_a\lambda_a = N_b\lambda_b = N_c\lambda_c = ...N_z\lambda_z \quad (1)$$

The N symbols refer to numbers of atoms, and $\lambda_a...\lambda_z$ are the radioactive decay constants. Equation (1) applies to a case of successive first-order radiodecay processes, i.e. a ==> b ===> c ... ===> z. Given sufficient time in a rigorously closed system, Equation (1) holds and a condition known as "secular equilibrium" exists. Many physical and chemical processes may bring about selective removal or addition of series members. The existence of disequilibria in the ^{238}U decay series has been recognized for some time [*1*] and is discussed in detail elsewhere [*2*]. The degree of disequilibrium is frequently expressed as an activity ratio (AR); i.e. the $^{234}U/^{238}U$ activity ratio is given as:

$$AR_{234/238} = N_{234}\lambda_{234} / N_{238}\lambda_{238} \quad (2)$$

Based upon Equation (2), when $AR_{234/238} = 1$, secular equilibrium exists, and the degree of disequilibrium is conveniently described using activity ratios. Disequilibria of $^{234}U/^{238}U$, $^{230}Th/^{234}U$, and other ^{238}U series member pairs are extensively used in the earth sciences to examine the formation of young volcanic rocks [*3*], sediment-water interactions [*4-5*], groundwater mixing [*6*] and as a means of radiometric age determination [*7*].

In principle, the chemical behavior of ^{234}U and ^{238}U ought to be indistinguishable; however, it is found that the $AR_{234/238}$ in some minerals is not unity due to α-recoil and/or selective leaching of ^{234}U from damaged lattice sites [*8*]. Hydrogeochemical processes commonly result in an aqueous phase that contains a slight excess of ^{234}U, and a solid (rock/soil) phase containing a deficiency of ^{234}U. This is typified by the $AR_{234/238}$ of 1.15 which is consistently observed for dissolved uranium in the oceans [*9-10*].

Measurements of $AR_{234/238}$ have traditionally been made by α-spectrometry. While relatively inexpensive instrumentation is involved, extensive chemical separations and source preparations are usually required, followed by long counting times (several hours-days) which brings about rather low sample throughput. In addition, sample sizes must be relatively large, i.e., about 1 μg or more of ^{238}U and equivalent activities of daughters are required. For natural samples, α-spectrometry cannot practically provide precisions of better than about 2% RSD, owing to the poor counting statistics for low-activity samples. The precision and sample size deficiencies of α-spectrometry are well-addressed by thermal ionization mass spectrometry (TIMS). High precision (< 0.5 % RSD) $^{234}U/^{238}U$ and $^{230}Th/^{234}U$ AR measurements have been made by TIMS [*11-16*]; these procedures are applicable to sample sizes of 10-100 ng ^{238}U. Thermal ionization procedures usually require a sample preparation yielding a chemically pure salt of the element of interest for efficient *in vacuo* solid-source ionization. However, TIMS generally requires lengthy chemical separations in order to produce a chemically pure sample aliquot for efficient ionization. Thus, extensive chemical separation is still required. Sample throughput, while improved over α-spectrometry, is still limited to a small number of samples per day. Finally, the high cost of commercial TIMS instrumentation can be an obstacle.

The routine application of $^{234}U/^{238}U$ disequilibria measurements to problems in the earth and environmental sciences is perhaps restricted by the limitations of α-spectrometry, and the consideration that TIMS is poorly suited for large-scale use in a diverse variety of laboratories with large numbers of samples. In contrast, we show that high-precision $AR_{234/238}$ measurements may be routinely obtained by inductively coupled plasma mass spectrometry (ICPMS), which has been widely developed for quantitative analysis and rapid isotopic measurements [*17*]. ICPMS is not expected to perform as well as TIMS for smaller samples, owing to the poorer atom utilization efficiency in the former (i.e., one in 10^4 - 10^7 atoms are counted in ICPMS as compared to one in 10^2-10^3 atoms in TIMS). Quadrupole ICPMS units also generally exhibit a much greater degree of mass discrimination (on the order of 1% relative per m/z) than the 0.1-0.2% relative per m/z which is prevalent in sector TIMS devices. Nevertheless, ICPMS offers less stringent sample preparation and operator skill requirements; it is routinely available in many laboratories worldwide, and can efficiently process large numbers of samples. It is not expected that quadrupole ICPMS will replace TIMS for critical geochronological measurements of $AR_{234/238}$; however, a need clearly exists for improvements upon the capabilities of α-spectrometry for high-volume routine measurements of $AR_{234/238}$ in natural waters.

The capabilities of ICPMS for U isotopic measurements have been recently exploited by several groups [*18-25*]; much interest exists in the determination of $^{235}U/^{238}U$. Crain and co-workers [*23*] compared α-spectrometric and ICPMS determinations of ^{230}Th, ^{239}Pu, and $AR_{234/238}$ in soil leachates; good agreement between the two techniques was found for $AR_{234/238}$, which was measured by ICPMS with a RSD of 3-5%. Small systematic differences were attributed to the use of different isotope dilution spikes. Hollenbach and co-workers [*25*] presented a scheme for rapid determinations of concentrations of the nuclides ^{230}Th and ^{234}U using on-line extraction chromatography and ICPMS.

We demonstrate herein that routine, high-precision determinations of $AR_{234/238}$ are feasible using ICPMS following straightforward sample preparative schemes. We have overcome difficulties associated with low abundances for ^{234}U (natural abundance 0.0055% at secular equilibrium), the low value for the $^{234}U/^{238}U$ atomic ratio (0.00005472 +/- 0.00000012 at secular equilibrium [*10*]) and the rather poor abundance sensitivities of quadrupole mass spectrometers. The latter problem may produce an overlap of ^{235}U at m/z 234 which amounts to as much as 1-2% of the ^{234}U signal. Schemes are presented for routine determination of $AR_{234/238}$ in natural waters and carbonates following suitable sample preparation via ion exchange, Fe(III) coprecipitation, and/or extraction chromatography. Since the $AR_{234/238}$ in the oceans is well-established through published α-spectrometric and TIMS studies, the $AR_{234/238}$ of seawater and contemporary carbonate precipitates serve as intrinsic, natural reference materials. Determinations of $AR_{234/238}$ are made by ICPMS through measurement of the $^{234}U/^{235}U$ atom ratio as a proxy for the $^{234}U/^{238}U$ atom ratio; mass discrimination is determined internally using within-run measurements of the ion ratio $^{191}Ir^{40}Ar^+/^{193}Ir^{40}Ar^+$ from an added Ir spike.

Experimental Method

Samples and Reagents

Surficial seawater was collected in HCl-cleaned high-density polyethylene containers from two Pacific Ocean shore locations at Carl G. Washburne Memorial State Park (Lane Co., OR) and at Kankaako Waterfront State Recreation Area (Honolulu, HI). Seawater samples were filtered using 0.45 μm capsule filters (Gelman 85217B, Gelman Sciences, Ann Arbor, MI); following filtration, 5 mL of 12 M HCl were added per L seawater. Bottled spring waters were obtained commercially and were used without filtration. Brands were as follows: Naya (Mirabel, Quebec, Canada), San Pellegrino, (San Pellegrino, Italy), and Crystal Geyser (Roxane Springs, Calistoga, CA, USA). Lake Erie surface water samples were collected from a shore location (Lake Co., OH, USA) using HCl-cleaned 19 L high-density polyethylene containers and were filtered and acid-preserved as described for seawater samples. Contemporary coral specimens, representing three species of various south Pacific origin, were obtained from an aquarium supplier, South Pacific Imports (Youngstown, OH). In selection of coral materials, specimens containing any visible detritus were avoided, as these phases may contribute significant non-carbonate uranium to the sample solution. In all cases, the acid-insoluble material (chiefly silicates) represented < 0.1 % of the sample mass. Bulk coral samples were broken into small pieces of 0.5 to 2 grams prior to dissolution. A sample of a fertilizer (Scotts Bulb Food, 19-26-6) was obtained commercially. Fossil mollusk shell fragments (*Unionacea sp.*) of 3 Ma age were obtained from J.L. Aronson, Case Western Reserve University. A U solution of natural $^{238}U/^{235}U$ isotopic composition was obtained from High Purity Standards, Charleston, SC. A carnotite U ore specimen (Montrose Co., CO, USA) was also used as a source of natural U. Reagent grade chemicals were used in dissolution, coprecipitation, ion exchange, and extraction chromatography; deionized water was produced in-house. Preparative operations were performed in disposable high-density polyethylene or polypropylene containers which were used as received. Sample handling and preparation steps were conducted in an ordinary laboratory environment; low chemical blanks of ~1% or less of the sample U content demonstrated that contamination during chemical processing was not significant.

Extraction Chromatography

Extraction chromatographic separations of U from carbonate (Ca-laden) or coprecipitated natural water (Fe-laden) matrices were performed using UTEVA resin (100-150 μm range, P/N UT-B25-A, EIChrom, Darien, IL). The UTEVA resin consists of dipentyl pentane phosphonate coated on a nonionic acrylic ester polymer support. The general properties of this material are detailed elsewhere [*26*]; U was retained in aqueous solutions containing 6-8 M HNO_3; 0.05 M aqueous ammonium oxalate was used for U elution. Two different low-pressure column configurations were used as dictated by sample solution volumes. Peristaltically pumped columns containing 0.5-1.0 g UTEVA resin in polyethyl ether ketone (PEEK) or fluorinated ethylene-propylene cartridges were used for sample solutions of 5-50 mL volume, while gravity-fed columns containing 1.5-

2.0 g UTEVA resin in borosilicate glass were used for sample solutions of 30-150 mL volume. After packing, the columns were conditioned using three cycles of 20 free column volumes (FCV) 8 M HNO_3 followed by 20 FCV of 0.05 M aqueous ammonium oxalate. The UTEVA columns were generally used for 20-30 analytical cycles before the material was discarded, although no significant change in performance was observed over this time.

Preparation of Carbonates

Specimens were dissolved in 8 M aqueous HNO_3; the solutions were filtered through 0.2 μm cellulose nitrate membrane filters to remove small amounts of undissolved material (chiefly silicates). For most analytical work, 50 mL of a solution containing 10 g dissolved coral was loaded onto a 0.6 g UTEVA column (initially washed with 20 FCV 8 M HNO_3) at a flowrate of 3-4 mL/minute. The column was washed with 20 FCV of 8 M HNO_3 in order to remove matrix elements; U was eluted using 5-10 mL of 0.05 M aqueous ammonium oxalate. The ammonium oxalate solution was collected and spiked with 200 mg/L Ir (derived from solid ammonium hexachloroiridate (III), Alfa Aesar). The chemical yields (measured using ^{238}U as an intrinsic yield tracer) were typically 90-95%; the unrecovered U was found to persist on the UTEVA column and was removed by two column rinse cycles with 15 mL ammonium oxalate solution followed by 15 mL of 8 M HNO_3.

Seawater and Bottled Water Preparation by Fe(III) Coprecipitation

Three-liter aliqouts of acidified sample were spiked with 150 mg Fe(III); the solution was heated to 80-90° C, then sparged for 5-10 minutes with nitrogen to expel any remaining CO_2. 15 M aqueous ammonia was added to achieve a pH of approximately 10. The Fe(III) precipitates representing one or more 3 L water samples were collected by decantation and filtration and combined; the composite precipitate was redissolved in a minimum volume (typically 20-30 mL) of 8 M HNO_3. The HNO_3 solution was processed by extraction chromatography (UTEVA resin) in the same manner as was done for dissolved carbonates. Overall chemical yields were in the 80-85% range using this procedure. The column was rinsed between samples as described above for carbonates in order to eliminate any cross-contamination.

Preparation of Lake Erie Surface Waters: Initial Preconcentration using a Weakly Acidic Cation Exchange Resin

The Fe(III) coprecipitation was impractical for Lake Erie water samples, owing to the low U concentrations of 0.25-0.30 μg/L which are prevalent. An initial preconcentration was therefore accomplished with a weakly acidic iminodiacetate polymeric resin (Chelex 20, BioRad). Prior to packing into columns, 50 g of this material was washed with 100 mL each of 1 M aqueous HCl, saturated aqueous ammonium acetate, and water. Columns were prepared in high-density polyethylene tubes (15 mm i.d. x 50 mm length); borosilicate glass wool was used as a bed support. 19 liter portions of water, adjusted to pH 4.0 +/- 0.1 with saturated ammonium acetate solution, were continuously circulated through the Chelex column for 6-8 hours at a flowrate of 0.5-1.0 liters/minute. Processing of 38 L of sample (containing 11 μg U) permitted final recovery of 8-9 μg U which made feasible the precise determination of $AR_{234/238}$ (*vide infra*). After Chelex 20 preconcentration, the U was eluted with 50 mL of 8 M HNO_3; this solution was then processed by extraction chromatography to produce 10 mL of a U-containing extract.

Preparation of Fertilizer and U Ore

The U ore sample was prepared for analysis by alkaline fusion with KOH in a Ni crucible, followed by dissolution of the melt in aqueous nitric acid. The extraction chromatography step was not needed, and analytical aliqout solutions containing 1-5 mg/L U were prepared by dilution. Fertilizer samples were prepared by dissolution in 8M HNO_3 followed by extraction chromatography with UTEVA resin columns using the scheme employed for carbonates.

Spectrochemical Measurements

ICPMS isotopic measurements were obtained using a Perkin-Elmer Sciex Elan Model 500 instrument operating in the peak-hop (either one or three points per mass spectral peak) mode. An active film multiplier (AF 561, ETP Scientific, Auburn, MA) was used in this instrument. With pneumatic nebulization, this particular ICPMS device produces maximum signal levels of about 5 x 10^6 ions/sec for a 1 mg/L U solution; in order to make measurements of ^{234}U signals, sample solutions contained approximately 0.5-5 mg/L total U. To optimally utilize the available solution volumes, the samples were analyzed using various laboratory-constructed externally mounted recirculating spraychambers; these were produced using 125 or 500 mL fluorinated ethylene-propylene bottles (Nalgene). These simple single-pass spraychambers consisted of a nebulizer inlet drilled into the cap, a drain hole at the bottom of the bottle, and an aerosol outlet drilled into the middle of the cylindrical portion of the bottle. The aerosol outlet was plumbed directly into the ICP torch using a 1 meter length of 1/4" internal diameter fluorinated ethylene-propylene tubing. Self-aspirating concentric glass nebulizers (VHG Labs, Manchester, NH, USA) with uptake rates of 1.0-1.5 mL/minute, were used. The spraychamber was disconnected from the torch and rinsed with dilute aqueous HNO_3 for decontamination between samples; this could be accomplished without extinguishing the plasma by interruption of the nebulizer Ar flow. Data collection for 90-180 minutes was feasible for 5 mL sample aliquots using the above arrangements. Plasma conditions were adjusted to optimize formation of Ir argides; the argide signal maximum coincided closely with the maximum U^+ signal. Signals were collected at m/z 231, 233, 234, and 235 with equal peak dwell times of 50 ms. Individual integrations of 5.1 minutes were collected, representing 75 seconds integration for each m/z. Repetitive integrations were collected until the sample solution was exhausted or until desired precision was achieved.

All signals were initially corrected for the deadtime (50 ns) of the multiplier-electronics arrangement. The (m-1/m) abundance sensitivity (AS) varied in the 10^{-5} to 10^{-4} range, being appreciably influenced by the ion lens settings; the abundance sensitivity was determined on a daily basis through measurements of the $^{238}U^+$ overlap at m/z 237. Subtractive corrections to the observed signals at m/z 234 were performed based upon the calculated $^{235}U^+$ overlap as follows:

$$AS = i237 / (i235*137.88) \quad (3)$$

$$i234_{corr} = i234_{raw} - (AS)*i235 \quad (4)$$

Chemical blanks were found to be minor (i.e. < 1 % of the U recovered from the sample) or insignificant; blanks were counted and their signals were subtracted at m/z 234 and 235. The contribution of $^{234}U^1H^+$ at m/z 235 was found to be negligible.

Special Safety Considerations

The quantities of naturally occurring α-emitters being handled are typically on the order of 10 pCi or less and represent no significant radioactivity hazard. Reasonable laboratory precautions are appropriate for avoiding skin contact, inhalation, or ingestion.

Results

Mass Discrimination Behavior

Previous workers [*27-30*] have noted that the degree of mass discrimination is readily predicted and corrected for using a "power law" approach:

$$R_{true} = R_{meas}(1 + C)^{\delta m} \quad (5)$$

where R_{true} = true value of an isotope ratio, R_{meas} = measured value, C = bias factor, and δm = mass difference. Several workers have found that the value of C may be predicted in the measurement of an unknown isotope ratio using the measured value of C found from a known isotope ratio of another element of similar mass which is added internally to each sample. This "internal standard correction" approach has been successfully conducted in ICPMS lead isotopic measurements using $^{205}Tl/^{203}Tl$ as the known internal ratio [*27-29*], while $^{71}Ga/^{69}Ga$ was used in zinc isotopic measurements [*30*]. In the present work, a similar approach is taken by measuring C and correcting $^{234}U/^{235}U$ based upon the $^{191}Ir/^{193}Ir$ measured from an Ir spike added to each sample. The latter ratio is conveniently measured using the $^{191}Ir^{40}Ar^+/^{193}Ir^{40}Ar^+$ signal ratio. The production of MAr^+ or "argide" ions in the argon ICP is well-known [*17*], and the existence of these ions would ordinarily be viewed as undesirable. The present work illustrates that Ir forms sufficient yields of argide ions to allow indirect measurement of $^{191}Ir/^{193}Ir$; the measured 231/233 signal ratio corresponds very closely to the $^{191}Ir/^{193}Ir$ ratio since the contribution of $^{193}Ir^{38}Ar^+$ at m/z 231 amounts to about 0.1 % of the $^{193}Ir^{40}Ar^+$. Under ideal conditions, solutions containing 200 mg/L added Ir produced signal levels of about 6000 and 10 000 ions/sec at m/z 231 and 233, respectively; for comparison, a 1 mg/L U solution ideally produced ~30 000-40 000 ions/sec at m/z 235. This method is advantageous as it allows measurement of a pair of ions which are very close to the m/z range of the ions of interest, and does not involve the use of synthetic radioisotopes such as ^{233}U or ^{236}U. Examples of ICP mass spectral scans of an Ir solution and an Ir-spiked modern coral extract are shown in Fig. 1. In addition to U elemental and oxide ions, a variety of polyatomic ions are evident; these ions consist principally of various isotopic forms of IrO_2^+, $IrAr^+$, $IrCl^+$, and IrO_3^+.

The measured 231/233 signal ratio is used to perform a power law mass discrimination correction of the following formulation:

$$({}^{234}U/{}^{235}U)_c = ({}^{234}U/{}^{235}U)_m(0.5949*i233/i231)^{0.5} \qquad (6)$$

where $({}^{234}U/{}^{235}U)_m$ is the measured isotope ratio, i231 and i233 are dead-time corrected signals at the respective masses, 0.5949 is the $^{191}Ir/^{193}Ir$ value [*31*] and $({}^{234}U/{}^{235}U)_c$ is the corrected isotope ratio. Under actual analytical conditions, i233/i231 was typically found to be in the 0.585-0.605 range. Results are reported herein as activity ratios as $AR_{234/238}$, calculated as follows:

$$AR_{234/238} = ({}^{234}U/{}^{235}U)_c \,/\, (137.88*0.00005472) \qquad (7)$$

The factor 137.88 is the naturally occurring $^{238}U/^{235}U$ atom ratio, and 0.00005472 is the $^{234}U/^{238}U$ atom ratio at secular equilibrium [*10*]. As the inclusion of the natural $^{238}U/^{235}U$ atomic ratio implies, this procedure assumes that all sample U is of natural $^{238}U/^{235}U$ composition. While some site-specific environments exist where this will not be the case, the vast majority of natural waters and geological materials will reflect the 137.88 value. Moreover, this assumption is also easily checked using a small portion of the sample U extract.

In order to investigate the efficacy of Equation (6), an experiment was performed in which $({}^{234}U/{}^{235}U)_m$ was systematically altered by changing the ELAN 500's ion lens settings. The results of this experiment, conducted using a U solution derived from a contemporary coral specimen (*Pocillopora meandrina*), are shown in Table 1.

TABLE 1--*Influence of ion optics settings upon Ir and U mass discrimination.*

B Lens, V	Measured $^{191}Ir/^{193}Ir$	Raw $AR_{234/238}$	Corrected $AR_{234/238}$
+1.23[1]	0.6017 +/- 0.0023	1.152 +/- 0.004	1.145 +/- 0.005 (n=15)[2]
+1.78	0.6006 +/- 0.0016	1.149 +/- 0.007	1.144 +/- 0.008 (n=12)
+2.23	0.5882 +/- 0.0008	1.141 +/- 0.004	1.148 +/- 0.004 (n=10)
+2.75	0.5931 +/- 0.0008	1.144 +/- 0.008	1.146 +/- 0.008 (n=10)

[1]The B lens was set at the indicated voltages; the potentials of the E1, P, and S2 lenses were held constant. [2]Uncertainties quoted are +/- 2σ standard error; values in parentheses indicate the number of replicates.

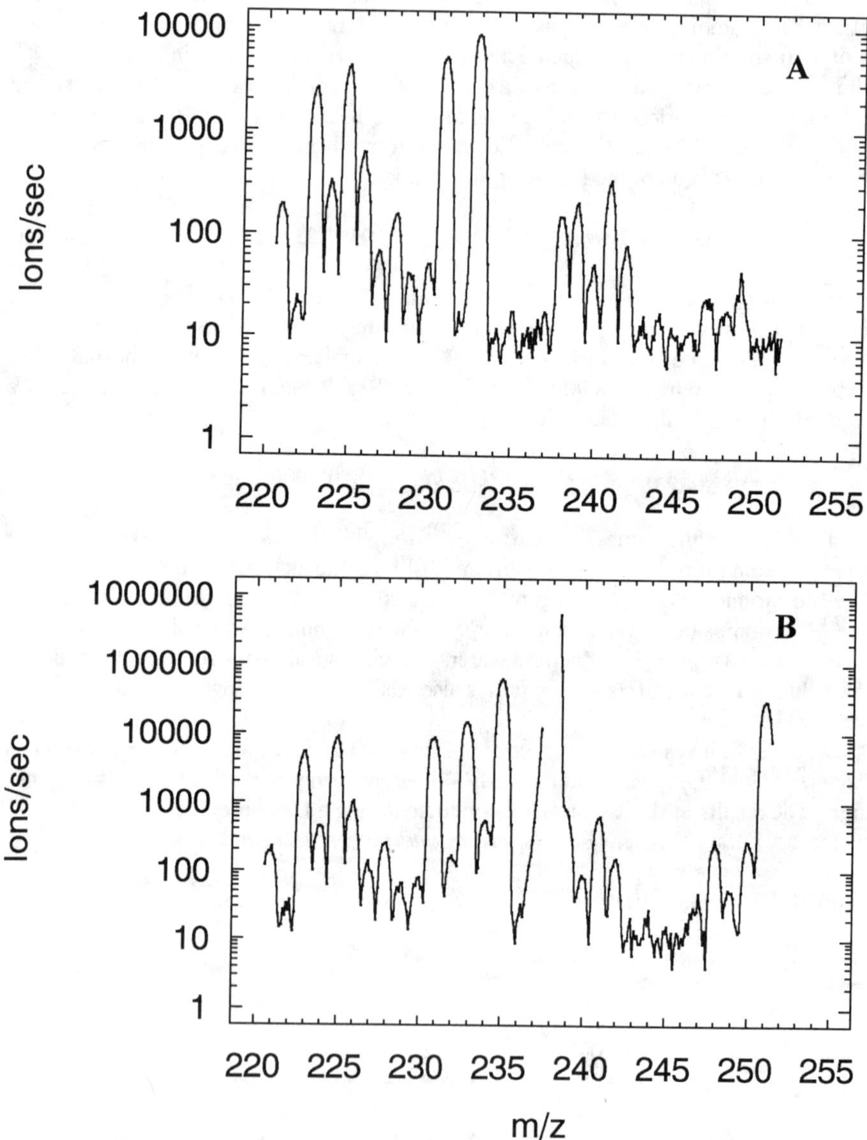

FIG. 1--*ICP mass spectra of (A) a blank solution containing 200 mg/L Ir, and (B) a U-containing extract prepared from a contemporary coral specimen. The portion of the* $^{238}U^{+}$ *peak between m/z 237.5 and 238.5 was not scanned in (B).*

The trends observed in this particular experiment show that adjustments of the B lens potential bring about systematic changes in $^{191}Ir/^{193}Ir$ and correlated changes in the uncorrected $AR_{234/238}$. The $AR_{234/238}$ values corrected via Equation (6) more closely resemble the expected value ($AR_{234/238}$ = 1.148 +/- 0.002 [*32*]) for seawater and contemporary carbonates. The Equation (6) correction is not large and is probably avoidable through careful adjustment of ion lens settings and routine monitoring of an isotopic standard; however, we have chosen to implement this correction rountinely to avoid possible biases due to unanticipated changes in mass bias factors.

Seawater Results

Table 2 depicts results for $AR_{234/238}$ in seawater obtained by the present ICPMS procedure, and comparative TIMS published values. The ICPMS results have been computed using Equations (6) and (7).

TABLE 2--*Comparison of ICPMS and published TIMS seawater results*[1].

Sample Location	Technique	$AR_{234/238}$
Oregon Coast # 1	ICPMS	1.146 +/- 0.004 (n=30)
Oregon Coast # 2	ICPMS	1.150 +/- 0.003 (n=27)
Oregon Coast # 3	ICPMS	1.154 +/- 0.005 (n=30)
Oahu, Hawaii	ICPMS	1.152 +/- 0.004 (n=21)
Pacific, Raft[2]	TIMS	1.140 +/- 0.005 (NR)[3]
Pacific, 30 m	TIMS	1.150 +/- 0.005
Pacific, 2000 m	TIMS	1.141 +/- 0.004
Pacific, 4900 m	TIMS	1.144 +/- 0.006
Atlantic, 10 m[4]	TIMS	1.145 +/- 0.007
Atlantic, 690 m	TIMS	1.140 +/- 0.007
Atlantic, 1640 m	TIMS	1.142 +/- 0.008
Atlantic, 2910 m	TIMS	1.142 +/- 0.007
Atlantic, 4280 m	TIMS	1.148 +/- 0.008

[1]Uncertainties quoted are +/- 2σ standard error for all ICPMS and TIMS results, values in parentheses indicate the number of replicates. [2,4]Samples are from locations as reported in Reference 10. [3]Numbers of replicates are not reported in Reference 10.

It is apparent that the ICPMS-derived results are in excellent agreement with the benchmark TIMS results; the sample averages and estimates for 2σ standard errors are both very similar. Both sets of results are obtained using similar counting times of 1-3 hours per sample. In terms of sample utilization, TIMS offers great advantages over our ICPMS procedures based upon the sample mass consumed. The ICPMS results were obtained using approximately 25 μg of U recovered from 9 L of seawater, while comparable TIMS results were obtained using 0.03 μg of U recovered from 10 mL of seawater. However, it is noted that the ELAN 500 ICPMS instrument used herein produces signal levels which are approximately 100X lower than are possible using presently available quadrupole ICPMS devices. Therefore, it is realistic to expect that results comparable to Table 2 may be obtainable via other ICPMS devices using < 0.1 L of sample, and possibly a less involved preconcentration strategy.

Modern Coral Results

Table 3 displays mass discrimination-corrected $AR_{234/238}$ results for contemporary corals produced by ICPMS, and illustrates their favorable comparison to published results from three TIMS laboratories. As is the case the comparison of Table 2, excellent agreement of sample averages is noted; the 2σ standard error values produced by ICPMS are comparable to values from References 32-34 but higher than the value from Reference 35; Stirling and co-workers [*35*] utilized multi-collector TIMS which provided improved precision over single-collector TIMS. The ICPMS-TIMS comparison in Table 3 again underscores the unrealistically large coral sample size (10 g) used in the present work. Regardless of this, Tables 2 and 3 clearly illustrate that highly precise, unbiased $AR_{234/238}$ values are readily obtained using ICPMS techniques. Using a modern ICPMS device which produces signal levels of 1-5 x 10^8 ions/sec at m/z 238 for 1 mg/L U, the methodology presented herein could clearly be adapted towards analysis of much smaller sample sizes than has been possible with the ELAN 500.

Results for Lake Erie and Bottled Water

Table 4 illustrates $AR_{234/238}$ for Lake Erie and bottled water samples. The experimental precisions reflect those observed for the contemporary coral and seawater samples. It is clear that $^{234}U/^{238}U$ disequilibrium exists for all of the samples reported below. This actually reflects the expected result for most natural waters, namely, a slight excess of ^{234}U is present above the ^{238}U-supported activity. A careful systematic study of $AR_{234/238}$ for Lake Erie waters is feasible using the methods reported herein; this information would be of use in apportioning riverine inputs and/or studying water/soil weathering interactions in the Great Lakes drainage basin. The major ionic contents of the bottled water samples are typical of Ca-Mg-bicarbonate groundwaters; this methodology could readily be applied to large-scale freshwater studies.

Results for Fertilizer and Fossil Mollusk Samples

In Table 5, results are shown for a N-P-K fertilizer and a *Unionacea sp.* fossil mollusk sample. The fertilizer sample yields an $AR_{234/238}$ value which is indistinguishable from unity, i.e., ^{234}U and ^{238}U are in secular equilibrium. Uranium is present in P-containing fertilizers owing to its association with authigenic marine phosphorite deposits. Given that most phosphorus deposits currently being mined are of Miocene age (12-26 Ma), secular equilibrium is expected for these materials. The *Unionacea sp.* fossil mollusk sample of 3 Ma age would be expected to reflect secular equilibrium of ^{234}U and ^{238}U, provided that the material truly existed in a closed-system environment. Since this is not found to be the case, one must conclude that open-system conditions have been prevalent at some point in the history of the sample. This finding indicates that the sample has been subjected to diagenetic effects leading to cation replacement. The ICPMS method is therefore useful for rapid testing of closed-system behavior for materials of independently established age.

Long-Term Repeatibility

The repeatibility of the $AR_{234/238}$ measurements was demonstrated using batchwise measurements of two U materials of natural $^{238}U/^{235}U$ isotopic composition (U ore and High Purity Standards U solution). Both of these materials exhibit $AR_{234/238}$ values close to, but slightly deviant from unity. Results are compiled in Table 6 for these

TABLE 3--*Comparison of ICPMS and published TIMS modern coral results.*

Sample Type	Technique	$AR_{234/238}$
Pocillopora meandrina #1	ICPMS	1.143 +/- 0.009 (n=10)[1]
Pocillopora meandrina #2	ICPMS	1.147 +/- 0.004 (n=30)
Pocillopora meandrina #3	ICPMS	1.150 +/- 0.007 (n=7)
Pocillopora meandrina #4	ICPMS	1.152 +/- 0.003 (n=44)
Pocillopora damicornis #1	ICPMS	1.149 +/- 0.006 (n=15)
Pocillopora damicornis #2	ICPMS	1.147 +/- 0.005 (n=20)
Pocillopora damicornis #3	ICPMS	1.144 +/- 0.006 (n=18)
Acropora cervicornis #1	ICPMS	1.150 +/- 0.002 (n=20)
Acropora cervicornis #2	ICPMS	1.152 +/- 0.004 (n=25)
1-Pc(U)[2]	TIMS	1.148 +/- 0.003 (NR)[3]
1-Pr1(U)[2]	TIMS	1.146 +/- 0.003
14-Pc(1)[2]	TIMS	1.149 +/- 0.003
TAN-E-1g[4]	TIMS	1.149 +/- 0.006
CWS-F-1[4]	TIMS	1.152 +/- 0.004
MS-63[5]	TIMS	1.147 +/- 0.007
Modern Coral[6]	TIMS	1.149 +/- 0.001

[1]Uncertainties reflect +/- 2σ standard error; n refers to the number of replicates.
[2]1-Pc(U), 14-Pr1(U), and 14-Pc(1) are recent corals reported in Reference 32.
[3]The numbers of replicates are not reported in References 32-35.
[4]Reference 33. [5]Reference 34. [6]Reference 35.

TABLE 4--*$AR_{234/238}$ results for Lake Erie and bottled water samples.*

Sample I.D.	U, µg/L	Volume Processed, L[1]	$AR_{234/238}$ Results[2]
Lake Erie #1	0.29	38	1.177 +/- 0.004 (n=15)
Lake Erie #2	0.29	38	1.164 +/- 0.003 (n=15)
Naya	0.60	12	1.093 +/- 0.005 (n=8)
San Pellegrino	7.9	6	1.115 +/- 0.003 (n=11)
Crystal Geyser	27	3	1.069 +/- 0.004 (n=17)

[1]The initial preconcentration of Lake Erie water samples was performed by ion exchange (Chelex 20). The initial preconcentration step of bottled water samples was performed by Fe(III) coprecipitation. [2]Uncertainties reflect +/- 2σ standard error; n refers to the number of replicates.

TABLE 5--*$AR_{234/238}$ results for fertilizer and fossil mollusk samples*

Sample I.D.	U, µg/g	$AR_{234/238}$ Results[1]
Scotts Bulb Food Fertilizer	83	1.003 +/- 0.006 (n=10)
Unionacea sp. fossil mollusk	9.2	0.933 +/- 0.008 (n=15)

[1]Uncertainties reflect +/- 2σ standard error; n refers to the number of replicates.

two materials; it is apparent that differences in the daily means of each material are not of any practical significance, and that the measurement method is capable of providing values which are repeatable over time.

Comparison of Experimental $AR_{234/238}$ Results vs. Counting Statistics Predictions

In Table 7, results of Poisson counting statistics calculations are presented for the experimentally measured ICPMS precisions based upon analyses presented in Tables 2 and 3. The predominant contribution to the counting statistics-predicted variance is the $^{234}U^+$ signal, which was counted at signal levels of 300-400 ions/sec. The comparison presented in Table 7 indicates that counting statistics constitute the chief source of imprecision in the measurement of $AR_{234/238}$ by the present method. This finding indicates the following: A) no significant additional source of variance, such as plasma flicker noise, is operative; B) over the timeframe of one analytical batch (1-3 hours) no significant drift in the mean $AR_{234/238}$ result is taking place; and C) further improvements in precision are possible if higher ^{234}U signals can be generated.

TABLE 6--*$AR_{234/238}$ results for uranium ore and High Purity Standards U stock.*

Sample I.D.	Analysis Date	$AR_{234/238}$ Results[1]
High Purity Standards	Mar 07 1997	0.977 +/- 0.004 (n=10)
High Purity Standards	Mar 10 1997	0.979 +/- 0.002 (n=40)
High Purity Standards	Mar 11 1997	0.981 +/- 0.002 (n=30)
High Purity Standards	Mar 12 1997	0.983 +/- 0.003 (n=20)
High Purity Standards	Mar 20 1997	0.980 +/- 0.003 (n=24)
High Purity Standards	Mar 25 1997	0.983 +/- 0.003 (n=25)
High Purity Standards	Mar 26 1997	0.977 +/- 0.004 (n=25)
Uranium Ore	Mar 10 1997	1.013 +/- 0.003 (n=36)
Uranium Ore	Mar 11 1997	1.014 +/- 0.004 (n=25)
Uranium Ore	Mar 12 1997	1.013 +/- 0.003 (n=30)
Uranium Ore	Mar 18 1997	1.013 +/- 0.003 (n=15)
Uranium Ore	Mar 19 1997	1.011 +/- 0.003 (n=15)
Uranium Ore	Mar 21 1997	1.015 +/- 0.004 (n=20)
Uranium Ore	Mar 22 1997	1.011 +/- 0.003 (n=15)

[1]Uncertainties reflect +/- 2σ standard error; n refers to the number of replicates.

TABLE 7--*Experimental vs. counting statistics predicted precision of $AR_{234/238}$.*

Sample Type[1]	Experimental % RSD	Predicted % RSD
Pocillopora meandrina #1	1.21 %	0.96 %
Pocillopora meandrina #2	0.97	0.98
Pocillopora meandrina #3	0.76	1.00
Pocillopora meandrina #4	0.85	0.80
Pocillopora damicornis #1	0.95	1.04
Pocillopora damicornis #2	0.90	0.81
Pocillopora damicornis #3	1.20	0.97
Acropora cervicornis #1	0.44	0.83
Acropora cervicornis #2	0.96	0.99
Oregon Coastal Seawater #1	0.84	0.77
Oregon Coastal Seawater #2	0.76	0.81
Oregon Coastal Seawater #3	1.24	1.14
Oahu, Hawaii Seawater	0.83	0.80

[1]The samples and locations correspond to the entries in Tables 2 and 3.

Conclusion

It has been demonstrated that precise $AR_{234/238}$ measurements may be readily made with quadrupole ICPMS. Values comparable to published TIMS results are attained for seawater and modern marine carbonate (coral) precipitates. The use of the method is demonstrated for a variety of natural waters and geologic materials. While the present work entailed using relatively large amounts (~5-25 μg) of U, this represents a limitation of the specific instrument used in this work and does not represent the ultimately expected capabilities of ICPMS. The procedures reported herein are potentially useful in studies of a diverse range of sample types, provided that a sufficiently concentrated U solution, with low dissolved solids content, can be isolated from the sample. The application of routine ICPMS methods to $AR_{234/238}$ measurements will potentially expand the applications of $^{234}U/^{238}U$ disequilibria toward a number of significant topics in aquatic and solid-phase geochemistry.

Acknowledgments

Portions of this work were supported by John Carroll University. The ICPMS instrument is a donation of Van Waters & Rogers Co. The authors also thank Charles A. Ramsey (EnviroStat, Inc.) and Timothy M.S. Gallagher (JCU) for providing seawater samples. James L. Aronson (CWRU) is thanked for providing the *Unionacea sp.* fossil specimen.

References

[*1*] Cherdyntsev, V.V., Orlov, D.P., Isabaev, E.A., Ivanov, V.I. *Geochemistry,* 1961, X, pp. 927-936.

[*2*] Ivanovich, M., Harmon, R.S. eds. *Uranium-series Disequilibrium: Applications to Earth, Marine, and Environmental Sciences,IInd Ed.*; Clarendon Press: Oxford, UK, 1992.

[*3*] Volpe, A.M., Hammond, P.E. *Earth and Planetary Science Letters,* 1991, CVII, pp. 475-486.

[*4*] Borole, D.V., Krishnaswami, S., Somayajulu, B.L.K. *Geochimica et Cosmochimica Acta,* 1982, XLVI, pp. 125-137.

[*5*] Scott, M.R. *Earth and Planetary Science Letters,* 1968, IV, pp. 245-252.

[*6*] Osmond, J.K., Rydell, H.S., Kaufman, M.I. *Science,* 1968, CLXII, pp. 997-999.

[*7*] Faure, G. *Principles of Isotope Geology IInd Ed;* John Wiley and Sons: New York, 1986.

[*8*] Dooley, J.R. Jr., Granfger, H.C., Rosholt, J.N. *Economic Geology,* 1966, LXI, pp. 1362-1382.

[*9*] Koide, M., Goldberg, E.D. in *Progress in Oceanography*; Sears, M., Ed.; Permagon Press: Oxford, U.K., Vol. 3, 1965, pp. 173-177.

[*10*] Chen, J.H., Edwards, R.L., Wasserburg, G.J. *Earth and Planetary Science Letters,* 1986, LXXX, pp. 241-251.

[*11*] Edwards, R.L., Chen, J.H., Ku, T.-L., Wasserburg, G.J. *Science,* 1987, CCXXXVI, pp. 1547-1553.

[*12*] Li, W.-X., Lundberg, J., Dickin, A.P., Ford, D.C., Schwarcz, H.P., McNutt, R., Williams, D. *Nature,* 1989, CCCXXXIX, pp. 534-536.

[*13*] Ludwig, K.R., Simmons, K.R., Szabo, B.J., Winograd, I.J., Landwehr, J.M., Riggs, A.C., Hoffman, R.J. *Science,* 1992, CCLVIII, pp. 284-287.

[*14*] Henderson, G.M.; Cohen, A.S.; O'Nions, R.K. *Earth and Planetary Science Letters,* 1993, CXV, pp. 65-73.

[*15*] McDermott, F., Grun, R., Stringer, C.B., Hawkesworth, C.J. *Nature,* 1993, CCCLXIII, pp. 252-255.

[*16*] Baker, A., Smart, P.L., Edwards, R.L., Richards, D.A. *Nature*, 1993, CCCLXIV, pp. 518-520.

[*17*] Jarvis, K.E., Gray, A.L., Houk, R.S. *Handbook of Inductively Coupled Plasma Mass Spectrometry*; Blackie: Glasgow, 1992.

[*18*] Russ, G.P., Bazan, J.M. In *Applications of ICPMS to Radionuclide Determinations, ASTM STP 1291*, American Society for Testing and Materials, 1995, pp. 131-140.

[*19*] Bolin, R.N. In *Applications of ICPMS to Radionuclide Determinations, ASTM STP 1291*, American Society for Testing and Materials, 1995, pp. 116-127.

[*20*] Walder, A.J., Hodgson, T. In *Applications of ICPMS to Radionuclide Determinations, ASTM STP 1291*, American Society for Testing and Materials 1995, pp. 20-25.

[*21*] Liezers, M., Tye, C.T., Mennie, D., Koller, D. In *Applications of ICPMS to Radionuclide Determinations, ASTM STP 1291*, American Society for Testing and Materials 1995, pp. 61-72.

[*22*] Manninen, P.K.G. *Journal of Radioanalytical and Nuclear Chemistry,* 1995, CCI, pp. 71-80.

[*23*] Crain, J.S., Smith, L.L., Yaeger, J.S., Alvarado, J.A. *Journal of Radioanalytical and Nuclear Chemistry,* 1995, CXCIV, pp. 133-139.

[*24*] Taylor, P.D.P., DeBievre, P., Walder, A.J., Entwistle, A. *Journal of Analytical Atomic Spectrometry,* 1995, X, pp. 395-398.

[*25*] Hollenbach, M., Grohs, J., Mamich, S., Kroft, M., Denoyer, E.R. *Journal of Analytical Atomic Spectrometry,* 1994, IX, pp. 927-933.

[*26*] Horwitz, E.P., Dietz, M.L., Chiarizia, R., Diamond, H., Essling, A.M., Graczyk, D. *Analytica Chimica Acta,* 1992, CCLXVI, pp. 25-37.

[*27*] Longerich, H.P., Fryer, B.J., Strong, D.F. *Spectrochimica Acta,* 1987, XLII, pp. 39-48.

[*28*] Ketterer, M.E., Peters, M.J., Tisdale, P.J. *Journal of Analytical Atomic Spectrometry,* 1991, VI, pp. 439-443.

[*29*] Ketterer, M.E. *Journal of Analytical Atomic Spectrometry,* 1992, VII, pp. 1125-1129.

[*30*] Roehl, R., Gomez, J., Woodhouse, L.R. *Journal of Analytical Atomic Spectrometry,* 1995, X, pp. 15-23.

[*31*] Barnes, I.L., Chang, T.L., DeBievre, P., Gramlich, J.W., Hageman, R.J.Ch., Holden, N.E., Murphy, T.J., Rosman, K.J.R., Shima, M. *Pure and Applied Chemistry,* 1991, LXIII, pp. 991-1002.

[*32*] Szabo, B.J., Ludwig, K.R., Muhs, D.R., Simmons, K.R. *Science,* 1994, CCLXVI, pp. 93-96.

[*33*] Edwards, R.L., Chen, J.H., Wasserburg, G.J. *Earth and Planetary Science Letters,* 1987, LXXXI, pp. 175-192.

[*34*] Stein, M., Wasserburg, G., Lajoie, K.R., Chen, J.H. *Geochimica et Cosmochimica Acta* 1991, LV, pp. 3709-3722.

[*35*] Stirling, C.H., Esat, T.M., McCulloch, M.T., Lambeck, K. *Earth and Planetary Science Letters,* 1995, CXXXV, pp. 115-130.

Ruth Hearn[1] and Heiko Wildner[1]

ISOTOPE RATIOS OF URANIUM USING HIGH RESOLUTION INDUCTIVELY COUPLED PLASMA - MASS SPECTROMETRY (ICP-MS)

REFERENCE: Hearn, R. and Wildner, H., "Isotope Ratios of Uranium Using High Resolution Inductively Coupled Plasma - Mass Spectrometry (ICP-MS)", *Applications of Inductively-Coupled Plasma-Mass Spectrometry to Radionuclide Determinations: Second Volume, ASTM STP 1344,* R.W. Morrow and J.S. Crain, Eds., American Society for Testing and Materials, 1998.

ABSTRACT: Actinide element isotope ratios have been determined in environmental samples using high resolution ICP-MS with ultrasonic nebulisation. Precisions as low as 0.1% RSD have been achieved using various methods of acquisition. The methodology has been used for environmental monitoring of uranium isotope ratios as an indicator of nuclear activity. Also, it has been applied to calcite dating studies as a measure of past geochemical disturbances.

KEYWORDS: high resolution ICP-MS, uranium, isotope ratios, environmental analysis

---

It is very important in environmental monitoring studies to be able to measure long-lived radionuclides. The measurement of uranium isotope ratios has to be accurate and precise in order to give meaningful information on the source of the uranium. Conventional radiometric techniques such as alpha spectrometry have limitations with respect to long-lived radionuclides due to relatively poor sensitivity, long counting times and chemical pre-treatment of samples.This leads to very low sample throughput and poor counting statistics.

[1] Senior Analysts, Analytical Services Group, AEA Technology Plc, B551 Harwell, Didcot, Oxfordshire, OX11 0RA England.

Quadrupole ICP-MS has been used to measure isotope ratios [1] but this method is also limited by poor precision and does not usually compete in terms of precision with more established techniques such as Thermal Ionisation Mass Spectrometry (TIMS). However, quadrupole ICP-MS does have some advantages over these techniques. Sample throughput is fast and extensive sample handling such as chemical separation is often not necessary.

Magnetic sector ICP-MS has been found to give better accuracy and precision on isotope ratio measurements by about an order of magnitude [2]. The advantage of this technique lies in the lower background (<0.1 counts per second (cps)), sensitivity of about 2000 cps/pg.g^{-1} when using ultrasonic nebulisation and the flat topped isotopic peaks when counting in low resolution mode. This allows uranium isotope ratio determination at pg.g^{-1} levels with satisfying count rates.

Magnetic sector ICP-MS can be used in high resolution mode to measure isotope ratios of lower mass elements such as Mg [3] and Cu [4]. However, in this paper, the authors have concentrated on the use of magnetic sector ICP-MS in low resolution mode ($M/\Delta M \cong 400$) for the determination of uranium isotope ratios in certified reference materials. Several different acquisition methods have been compared.

Instrumentation

The magnetic sector ICP-MS - the Plasma Trace 2 (Micromass UK Limited, Floats Road, Wythenshawe, Manchester M23 9LZ UK) - consists of an ICP, a sampling interface, a magnetic sector, an electrostatic analyser and the detectors. The instrument has two types of detector - an electron multiplier and a Faraday detector. For the purposes of the experiments described below, only the multiplier was used to collect data.

Sample introduction to the ICP was performed using either a meinhard nebuliser (Scitek, Newport Pagnell, UK) or an ultrasonic nebuliser with membrane desolvator USN-6000 (Cetac, 5600 South 42nd Street, Omaha, NE 68107, USA) in order to improve the efficiency about 20-fold relative to meinhard nebulisation. The instrument operating conditions are detailed in Table 1.

Reagents

Two certified uranium isotope ratio standards were used for this study - NBS050 and NBS200. These are certified for the following ratios - 234:238, 235:238 and 236:238. These solutions were diluted in a 2% nitric acid solution using ultra-pure nitric acid (Romil Ltd, Cambridge, UK.) and deionised water with a resistivity of 18.2 MΩ cm.

TABLE 1 -- Instrument operating conditions used with ultrasonic nebuliser

Coolant gas flow rate	14 L / min
Auxillary gas flow rate	0.5 L / min
Carrier gas flow rate	0.9 L / min
Sweep gas flow rate	3.0 L / min
RF forward power	1350 W

Experimental

Acquisition was performed by three different methods detailed below. Each method was tested with the reference materials at exactly the same concentration and with the same total acquisition time in order to allow direct comparison of performance. Instrument conditions were not varied between the methods.

Method 1 - Separate Window Acquisition

Each isotope is acquired individually in its own mass window. The lowest mass isotope is scanned first then the magnet "jumps" to the next isotope. Each time the magnet jumps, it needs time to settle (0.1-0.5 seconds). This settle time time can be reduced through the software but it still makes this method slow relative to the other two. The mass calibration has to be exact in order that the whole peak can be seen in each window. Each isotope can be scanned several times before jumping to the next - which can help to average out the noise on the peak. This method also allows the use of different dwell times or scan replicates for each individual isotope so that most of the total acquisition time can be spent on minor isotopes to improve the counting statistics and therefore improve the precision. In this experiment, method 1 was run twice, once with the same parameters for each isotope (1A) and once with a smaller dwell time on U-238 (1B). Method 1B is ideal for measurement of 234 : 238 ratios in calcite dating studies where the 234 peak is so much smaller than the 238 and the time spent on 234 has to be optimised in order improve counting statistics.

Method 2 - Scanning

The element menu for this method consists only of the central isotope (in this case U-236) but with a wide scan acquisition window such that all of the isotopes of interest are included. Unlike method 1, this does not allow for different acquisition parameters on each isotope. However, the mass calibration does not have to be as exact. This method is best suited to isotopes of similar abundance such as the three major lead isotopes but it can also work well for uranium isotope ratios.

Figure 1 shows an example of a scan on a sample containing uranium. Note the tailing of the 238 peak. This is typical of the abundance sensitivity of 10^5 that is found in this mass spectrometer.

Method 3 - Sprint Scan

This is similar to method 2 above in that the mass region containing all the isotopes is scanned and the same parameters have to be used on all isotopes in the scan. There are two main differences :

- several very fast scans are performed and are summed to give a final result
- only certain regions within the mass range are acquired. This means that no time is wasted acquiring data that is not used for calculation.

This method tends to give the best precision because the summing of the fast scans average out the noise from the plasma. Figure 2 shows an example of a sprint scan across the uranium isotopes. Note that there appears to be a distinct peak at 237. This is not really a peak, but rather the tailing of 238 which was clearly seen from the complete scan (method 2).

The measurement parameters for each of the four methods detailed above are given in Table 2. Method 1A, 1B, 2 and 3 were tested against each other by running two certified uranium isotope ratio standards, measured at exactly the same concentration. Five replicate runs were made at each method with a total acquisition time of five minutes per method. These methods were run several times with both of the certified standards and also with environmental samples with mostly natural isotope ratios.

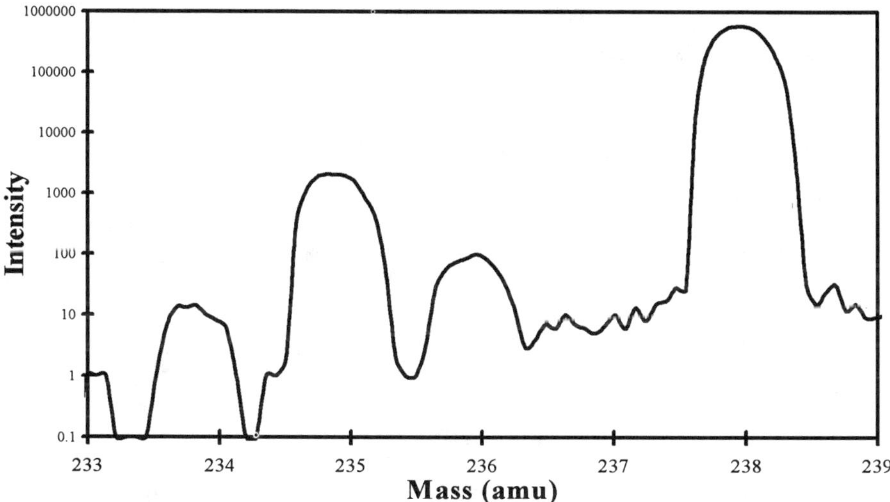

FIG. 1 -- Scan of a sample containing 500 pg/ml uranium with a 0.08 pg/ml addition of U-236. The isotope ratios are 234/238 = 0.0024% and 235/238 = 0.355%.

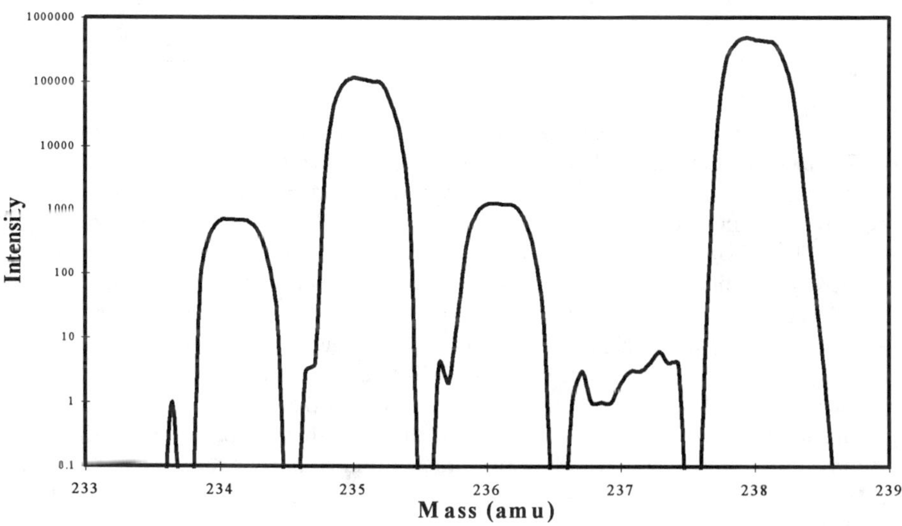

FIG. 2 -- Sprint scan of a sample containing 500 pg/ml uranium with a 0.08 pg/ml addition of U-236. The isotope ratios are 234/238 = 0.0024% and 235/238 = 0.355%.

TABLE 2 -- Measurement parameters for each of the four acquisition methods.

Parameter	Method 1A	Method 1B	Method 2	Method 3
234 Dwell Time (ms/point)	20	20	1	2
235 Dwell Time (ms/point)	20	20	1	2
236 Dwell Time (ms/point)	20	20	1	2
238 Dwell Time (ms/point)	20	1	1	2
Peak Widths	1.3	1.3	8	1.3
Number of sweeps	20	20	200	200
Time per run (seconds)	60	60	60	60

Results

Table 3 shows the average 235:238 ratios for the two certified standards with each of the 4 methods together with the expected results. Table 4 gives the range of precisions found with five replicate runs on the two standards with each of the three methods. Measurements on a sample containing an addition of U-236 gave a 236/238 ratio precision of 0.1%. This shows that when the isotope peaks are of a more similar size, precisions even lower than those shown in Table 4 are obtainable by the methods described.

These results show that in terms of bias, there is no significant difference between the methods. The sprint scan, however, does appear to give the best precisions. Results for each of the methods have a negative bias of between 4.4 and 8.4%. This type of bias is not uncommon and is due in part to selective sampling of the heavier isotope through the instrument interface. Xie and Kerrich [5] also found bias showing preferential reduction in response to the lighter isotopes when measuring Zr and Hf isotope ratios by quadrupole ICP-MS. The mass bias for Zr varied from -4% to -12%. Walder et al [6] also found negative bias for uranium 235/236 ratios.

The accuracy data also suggest that using different dwell times on each isotope (Method 1B) does not alter the accuracy of the isotope ratio measurement. Subsequent experiments outside the scope of this paper have confirmed this conclusion.

TABLE 3 -- Average 235/238 ratios for NBS050 and NBS200.

Method	NBS050	NBS200
1A	0.0505	0.236
1B	0.0491	0.230
2	0.0497	0.231
3	0.0503	0.234
Expected	0.0528	0.251

TABLE 4 -- Range of precisions on 235/238 ratios for NBS050 and NBS200.

Method	Precision - RSD%
1A	0.59 - 1.1
1B	0.36 - 1.4
2	0.31 - 1.4
3	0.27 - 0.86

Sprint Scan

From the results above, it can be seen that the sprint scan gives the best precisions. Therefore, an experiment was performed to test the sprint scan method more throughly. Three sprint scan experiments were set up each with differing measurement parameters. The peak widths for each of the three scans remained constant. These parameters are detailed in Table 5. Each sprint scan was replicated 10 times using a depleted uranium standard with about 10 ppb U-238. This standard was spiked with an additional amount of U-236 to increase the counting statistics for the 236/238 ratio.

TABLE 5 -- Measurement parameters for three sprint scans.

Parameters	Sprint 1	Sprint 2	Sprint 3
Dwell time (ms/point)	2	2	4
Sweeps	200	100	100
Time per run (s)	80	40	80

Sprint Scan Results

The accuracy of the measurements were comparable with those found above with the same negative bias. The precisions are detailed in Table 6.

TABLE 6 -- Precisions obtained with three sprint scans.

	Precision - % RSD		
Isotope Ratio	Sprint 1	Sprint 2	Sprint 3
234/238	11	16	12
235/238	1.1	1.8	0.99
236/238	4.8	6.3	3.8

This sprint scan experiment was performed using meinhard nebulisation (unlike the previous experiment on USN) so the precision on the low abundance isotopes (especially 234) was expected to be worse than before simply due to counting statistics. However, these results do give a good comparison of the measurement parameters. For all three ratios, sprint 2 gives the poorest precision. This should be of no surprise since this method has a shorter total acquisition time than the other two. Therefore, fewer counts per isotope are acquired and the method is more limited by counting statistics.

Sprint 3 appears to give the best precisions for 235/238 and 236/238. This suggests that longer dwell times on each isotope gives better precision than more sweeps. However, this is a very tentative conclusion as there were relatively few data points and the difference between sprint 1 and 3 is only very small.

Conclusions

This paper shows that high resolution ICP-MS can be used to successfully measure uranium isotope ratios to precisions as low as 0.27% on the 235/238 ratio. There are several methods by which the data can be acquired. The method which gives the best precision of those tested is the sprint scan. Determination of uranium isotope ratios with high precision can be used for the dating of geological samples as well as for the application of isotope dilution as a means of calibration [6].

Further improvements can be made using high resolution ICP-MS with multi-collectors [7]. This allows the measurement of each of the isotopes of interest simultaneously, thus eliminating the effect of fluctuations from the plasma.

Acknowledgements

Many thanks to Dr. J. Toole, Radionuclide Analysis, AEA Technology, for his interest in this project. Financial support by the Deutsche Forschungsgemeinschaft is gratefully acknowledged.

References

[1] Boorn, A., Fulford, J., Douglas, D. and Quan, E., "Isotope Ratio Measurements by ICP-MS," *ICP Information Newsletter*, Vol. 11. No. 10, April 1986 p. 783.

[2] Kim, C., Seki, R., Morita, S., Yamasaki, S., Tsumura, A., Takaku, Y., Igarashi, Y. and Yamamoto, M., "Application of High Resolution Inductively Coupled Plasma Mass Spectrometer to the Measurement of Long Lived Radionuclides," *Journal of Analytical Atomic Spectrometry*, Vol. 6, April 1991, p. 205.

[3] Vanhaecke, F., Moens, L., Dams, R. and Taylor, P., "Precise Measurement of Isotope Ratios with a Double-Focusing Magnetic Sector ICP Mass Spectrometer," *Analytical Chemistry*, Vol. 68, No. 3, February 1996, p. 567.

[4] Vanhaecke, F., Moens, L., Dams, R., Papadakis, I. and Taylor, P., "Applicability of High Resolution ICP - Mass Spectrometry for Isotope Ratio Measurements," *Analytical Chemistry*, Vol. 69, No. 2, January 15, 1997 p. 268.

[5] Xie, Q. and Kerrich, R., "Optimisation of Operating Conditions for Improved Precision of Zirconium and Hafnium Isotope Ratio Measruement by Inductively Coupled Plasma Mass Spectrometry (ICP-MS)," *Journal of Analytical Atomic Spectrometry*, Vol. 10, No. 2, February 1995, p. 99.

[6] Wollenweber, D., Wildner, H. and Wünsch, G., "Uranium determination in process chemicals: on the way to sub-pg/g concentrations," *Fresenius Journal of Analytical Chemistry*, Vol. 359, 1997, p. 414.

[7] Waldner, A.J., Koller, D., Reed, N.M., Hutton, R.C. and Freeman, P.A., "Isotope Ratio Measurement by Inductively Coupled Plasma Multiple Collector Mass Spectrometry Incorporating a High Efficiency Nebulization System," *Journal of Analytical Atomic Spectrometry*, Vol. 8, No. 7, October 1993, p. 1037.